Sorption - From Fundamentals to Applications

Edited by George Z. Kyzas

Published in London, United Kingdom

Sorption - From Fundamentals to Applications
http://dx.doi.org/10.5772/intechopen.98007
Edited by George Z. Kyzas

Contributors
Issoufou Amadou, José María Rivera, Guadalupe Vivar-Vera, Lidia E. Chiñas-Rojas, Yafeth F. Cruz-Martínez, , Seth Limón Colohua, Eric Houbron, Miguel Manuel Pérez Hernández, Isabel Cristina Páez-Pumar Romer, Isabella Victoria Plazola Santana, Rosa María Rodríguez Bengoechea, Rakesh Govind, Ankurman Shrestha, Joseph Munene Mwaniki, Naoki Kano, Zou Ming, David Eva Vanessa Anak, Muhammad Nabil Md Sari

First published in London, United Kingdom, 2022 by IntechOpen
IntechOpen is the global imprint of INTECHOPEN LIMITED, registered in England and Wales, registration number: 11086078, 5 Princes Gate Court, London, SW7 2QJ, United Kingdom

British Library Cataloguing-in-Publication Data
A catalogue record for this book is available from the British Library

Additional hard and PDF copies can be obtained from orders@intechopen.com

Sorption - From Fundamentals to Applications
Edited by George Z. Kyzas
p. cm.
Print ISBN 978-1-80355-969-8
Online ISBN 978-1-80355-970-4
eBook (PDF) ISBN 978-1-80355-971-1

Meet the editor

Dr. George Z. Kyzas is a professor in the Department of Chemistry, International Hellenic University (IHU), Greece. He studied Chemistry at the Department of Chemistry, Aristotle University of Thessaloniki (AUTh), Greece. He obtained his BSc, MSc, and Ph.D. in Chemical Technology from the same university where he also worked as a postdoctoral researcher. He is now working at the Department of Chemistry, International Hellenic University, Kavala, Greece, where he has been the head of the department since 2019. He is also the Director/Chair of the MSc in Cosmetic Chemistry. His research interests are the synthesis and characterization of various (majorly adsorbent) materials (inorganic, aluminates, polymers, graphenes, activated carbons, agro-food residues, nanomaterials, carbon nanotubes, etc.) for environmental applications (wastewaters treatment). Dr. Kyzas has published more than 220 scientific papers in international journals, 8 books, and 38 book chapters. He also holds three patents. He has been a guest editor for special journal issues and is the editor of Environmental Science and Pollution Research and is a reviewer for more than 200 other scientific journals. He has more than 120 announcements (invited) at international conferences. He was named in Stanford University's list of the World's Top 2% Scientists for 2019 and 2020. He has been awarded scholarships from the Research Committee of the Aristotle University of Thessaloniki (2009, 2013), the Greek State Scholarship Foundation (2013), and the Stavros Niarchos Foundation (2016). He has also participated in about twenty-five research projects.

Contents

Preface

Sorption - From Fundamentals to Applications describes various aspects of the sorption technique. Sorption is a physical and chemical process by which one substance becomes attached to another. Specific cases of sorption include: (1) absorption, which is the incorporation of a substance in one state into another of a different state (e.g., liquids being absorbed by a solid or gases being absorbed by a liquid); (2) adsorption, which is the physical adherence or bonding of ions and molecules onto the surface of another phase (e.g., reagents adsorbed to a solid catalyst surface); and (3) ion-exchange, which is an exchange of ions between two electrolytes or between an electrolyte solution and a complex. At the beginning of the 21st century, separation processes presented a comprehensive application of the major operations performed by various industries, such as chemical, food, environmental, and biotechnology industries. Sorption, which is one of the preferred separation processes because of its effectiveness at different interfaces, has attracted the attention of many scientists. This book presents a comprehensive overview of sorption and its various applications, as well as discusses its functions and importance in chemical and biochemical plants, including environmental treatment. It is a useful reference for those who are interested in sorption techniques in various industries.

I am grateful to all the authors who have contributed their tremendous expertise to this project. I also wish to acknowledge IntechOpen publishing for their outstanding support throughout the publication of this book.

George Z. Kyzas (MSc, PhD)
Professor,
Department of Chemistry,
International Hellenic University,
Kavala, Greece

Acknowledgments

We acknowledge the support we received for this edited book from the Greek Ministry of Development and Investments (General Secretariat for Research and Technology) through the research project "Intergovernmental International Scientific and Technological Innovation-Cooperation. Joint declaration of Science and Technology Cooperation between China and Greece" with the topic "Development of monitoring and removal strategies of emerging micro-pollutants in wastewaters" (grant No. T7DKI-00220).

Section 1

Sorption in Biomass

Chapter 1

Sorption of Pollutants in Wastewater Solids

Rakesh Govind and Ankurman Shrestha

Abstract

Sorption in wastewater solids is an important removal mechanism for pollutants in biological treatment systems. It is often an overlooked mechanism, since traditionally, excess solids from biological treatment were land filled. However, with the emergence of using wastewater solids as a potential fertilizer, pollutants sorbed into the solids can re-emerge as soil pollutants, with potential uptake by crops, and even transported into groundwater. This is especially applicable for hydrophobic chemicals, like alkyl fluorinated compounds (PFAS, PFOS), which have recently received widespread attention as pollutants in water bodies across the globe. In this chapter, sorption of pollutants in wastewater solids has been presented from both a thermodynamic analysis, involving equilibrium parameters, as well as a kinetic process involving transport to the cell walls and permeation through the cell membranes. Based on experimental data and models it is shown that biodegradation in wastewater systems is actually mass transfer coefficient for diffusive transport across the microbial cell walls.

Keywords: sorption, biomass, cell, permeability, transport, pollutants

1. Introduction

Transport of pollutants into wastewater solids (biomass) can occur due to a variety of mechanisms, some of which are surface related, while others involve absorption or sorption into the biomass cells. Uptake of pollutants by the surface of the biomass is defined as "adsorption" driven by several possible mechanisms which includes hydrophobic-hydrophobic interaction, electrostatic interaction and hydrogen bonding. The latter two mechanisms are dominant under basic and acidic conditions, respectively, while hydrophobic-hydrophobic interaction occurs mainly under neutral conditions [1].

Sorption or absorption in biomass involves the transport of the pollutants into the cell, rather than adsorbed on the outside surface of the cells. Sorption in wastewater solids is an important mechanism for removal of organic compounds in biological wastewater treatment systems. Experimental results and theoretical developments related to the sorption process have been reported in our previous work [2, 3]. During isotherm measurements, biological activity in wastewater solids, especially in activated sludge, that contains a substantial amount of active biomass, must be controlled to obtain accurate sorption isotherms. Measurements of varying aqueous phase concentrations could include the effect of biodegradation, resulting in apparent sorption capacities greater than actually achieved.

Sorption in Biomass can be analyzed as an equilibrium process, using isotherms, as well as a kinetic process, involving transport to the biomass cell walls and permeation through the cell membranes.

2. Analysis of equilibrium sorption: surface adsorption vs. partitioning

A dual-process mechanism [4] was postulated for the sorption of lindane and hexachlorocyclohexane on five unidentified bacteria. In this mechanism, sorption was viewed as a combination of rapid surface adsorption and diffusive penetration. A similar mechanism was suggested [5] for the uptake of 2,4-dichlorophenoxyacetic acid by *Pseudomonas fluoresces*. The result was a combination of the following steps: adsorption onto the cell wall accompanied by passive diffusion into the cytoplasm. Based on the comparable sorption results of organic compounds between live and dead biomass in a Kraft mill generated lagoon, it was suggested that sorption on biomass was mainly a passive diffusion of small molecules into the cell than active or facilitated transport [6].

By comparing the sorption of chloroethanes on different types of microbial biomass, it was speculated [7] that the sorption of organic molecules on microbial biomass was related to the leachable organic carbon which was released from the cell cytoplasm due to the rupture of cells. It was found that the more leachable organic carbon the biomass contained, the higher the uptake capacity of the toxic organic compound it had.

In order to better understand the sorption mechanism on microbial biomass, Bell [8–11] examined the magnitude of the heats of sorption for diazinon and lindane. The thermodynamic data strongly suggested a physical sorption instead of a chemical sorption. By reviewing the sorption mechanism on biomass in the literature, Bell [9] concluded that firm, general conclusions cannot be reached concerning the mechanism of sorption on biomass. Indeed, the mechanisms may be specific to the particular system of chemicals and biomass. It seems likely that a combination of adsorption and absorption may be responsible for the sorption on biomass and that the relative importance of each mechanism may vary from system to system [12].

Biological activity in wastewater solids, especially in activated sludge which contains a substantial amount of active biomass, must be controlled in order to obtain accurate sorption isotherms. Otherwise, measurements of changing aqueous phase compound concentrations could include the effect of biodegradation, resulting in apparent sorption capacities greater than actually achieved. Despite the difficulties associated with the control of the biological activity in biomass, several attempts have been reported.

In this chapter, sorption in biomass will be modeled as an adsorption partitioning process, in which physical adsorption on the surface and partitioning into the biomass cells occurs simultaneously. Wastewater solids contain high percentage (50–85%) of organic matter, in a relatively loose physical form. It is generally believed that the organic matter is a complex mixture of live and dead microorganisms, which are primarily proteins, fats and carbohydrates, and other organic sediments. While the organic matter is basically solid, it is certainly different from activated carbon. The nature of the organic matter gives the sludge certain characteristics of an organic solvent. Chiou et al. [13–15] described the sorption of toxic organic compounds on soil organic matter as a partitioning phenomenon. Using the Flory-Huggins theory, they were able to derive an equation relating the partition coefficient of a toxic organic compound between the aqueous phase and the soil organic matter to the molar water solubility of the compound. Essentially, the uptake of toxic organic compounds by sludge through dissolution into cells can be viewed as a partitioning process in which the toxic organic compounds are distributed between two phases: the aqueous solution and the organic matter of the cell.

In the early stage of the sorption process, the cell is basically free of the toxic organic compounds except for the amount adsorbed on the surface. The difference in chemical potential causes the toxic organic compounds to transport from the bulk liquid to the region close to the surface and to be adsorbed onto the surface. In the meantime, the partitioning and subsequent penetration (diffusion) of the toxic organic compounds into cells occurs in parallel with adsorption. When a molecule approaches the site already occupied by another molecule in a process to form a double layer on the sludge surface, unlike in the case of activated carbon adsorption in which the penetration of the adsorbate into the solid wall of carbon normally does not occur, the resulting higher chemical potential on this site is likely to cause the adsorbed molecule to be dissolved in the organic matter of the cell.

To test this hypothesis, isotherm measurements were conducted for the first four compounds in **Table 1** using pasteurized sludge to eliminate biodegradation. For the other compounds, listed in **Table 1**, untreated sludge was used, since these compounds are non-biodegradable within the 6-hour equilibriation time period. The model coefficients obtained for these compounds are also listed in **Table 1**, along with the average percentage error between the calculated and experimental points. **Table 2** lists the contribution of partitioning to the overall sorption amount (q_{pi}/q) along with the partition coefficient, k_{pi} and the octanol–water partition coefficient (K_{OW}).

For the last four compounds in **Table 2**, adsorption is negligible with over 90% sorption occurring due to partitioning. This is also consistent with the K_{OW} values for these compounds, which is an indication of the compound's hydrophobicity, and thus has a higher tendency to partition with the organic matter of the biomass. This shows that a compound's equilibrium partitioning into biomass due to sorption is correlated

Pollutant	Partition coefficient (k_p) l/g	Maximum adsorption capacity (q_a^o) mg/g	Equilibrium adsorption constant (k) l/mg	% Error
Methylene chloride	0.012	0.250	0.30	12.2
Chloroform	0.035	0.149	0.48	6.4
1,1-dichloroethylene	0.140	0.042	0.48	6.4
Carbon tetrachloride	0.210	0.100	1.40	21.5
Chlorobenzene	0.243	0.010	2.60	18.9
Tetrachloroethylene	0.653	0.012	5.29	9.7
Phenanthrene	6.8	0.020	1.29	25.0
Dibutyl phthalate	7.00	0.014	5.30	14.7

Table 1.
Best-fit parameters of the adsorption-partition (A-P) model for sorption of selected pollutants on activated sludge. % error is the average error between the experimental value of pollutant uptake and calculated value.

Pollutant	Partition coefficient (k_p) l/g	Octanol–water partition coefficient (K_{OW})	Total sorption (q) mg/g	q_p/q
Methylene chloride	0.012	18	0.07	0.172
Chloroform	0.035	93	0.083	0.420
1,1-dichloroethylene	0.140	134	0.156	0.899
Carbon tetrachloride	0.210	436	0.268	0.783
Chlorobenzene	0.243	690	0.250	0.971
Tetrachloroethylene	0.653	759	0.663	0.985
Phenanthrene	6.8	28,184	6.811	0.998
Dibutyl phthalate	7.00	158,489	7.012	0.998

Table 2.
Partition coefficient, octanol–water partition coefficient, Total sorption and ratio of amount partitioned to the Total sorption amount for selected pollutants in activated sludge. Liquid phase concentration of each pollutant is 1 mg/l.

directly with its K_{OW} value, and this correlation (correlation coefficient = 0.97) is given as follows:

$$\mathrm{Log}k_p = 0.73\mathrm{Log}K_{OW} - 2.64 \quad (1)$$

Experimental studies were conducted to better understand the competition effects in multicomponent systems. it was found that compounds which primarily adsorb on the surface have a strong impact of competitive adsorption, with the adsorption extent of one compound varying greatly with the concentration of the second compound. This is due to the fact that the number of surface adsorption sites on biomass are limited and adsorption of one compound will impact the adsorption capacity of the second compound.

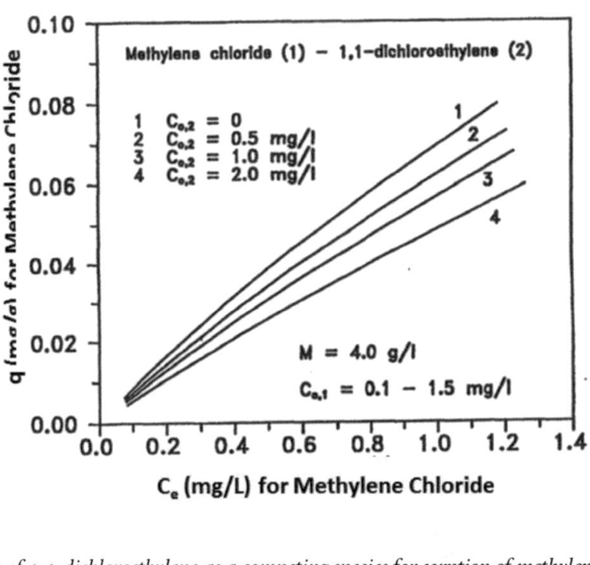

Figure 1.
Effect of concentration of 1,1-dichloroethylene as a competing species for sorption of methylene chloride.

However, for compounds which primarily undergo sorption instead of surface adsorption, there is no competitive sorption between the compounds. The adsorption-partitioning (A-P) model was used to fit the sorption of methylene chloride, chloroform and tetrachloroethylene in the presence of 1,1-dichloroethylene. According to **Table 2**, the percentage of partitioning in the overall uptake for these compounds are 17.2%, 42% and 98.5%, respectively. Hence, the effect of competition would be most significant for methylene chloride and least significant for tetrachloroethylene. The simulation results are shown in **Figures 1–3**. For tetrachloroethylene, there is almost no competition effect, as was the case where chlorobenzene was the competing

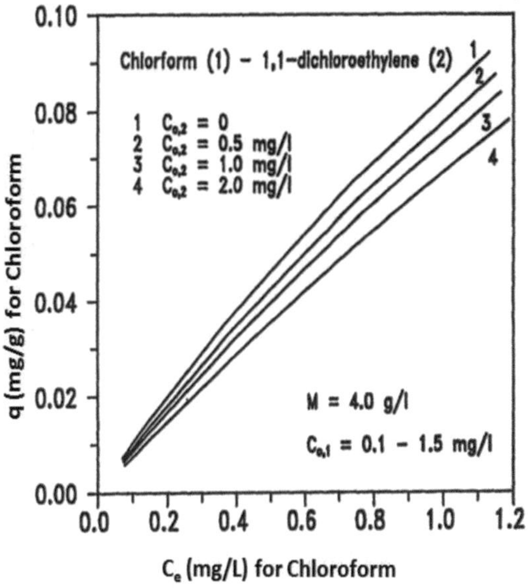

Figure 2.
Effect of the concentration of 1,1-dichloroethylene as a competing species for sorption of chloroform.

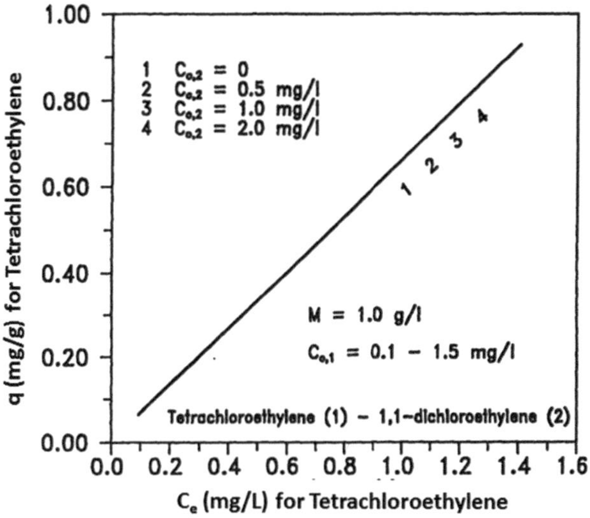

Figure 3.
Effect of the concentration of 1,1-dichloroethylene as a competing species for sorption of tetrachloroethylene.

Key pollutant	Competing pollutants	Competing effect	K_{OW}	Reference
Phenol (P)	2-chlorophenol (CP), 2-Nitrophenol (NP)	Yes Yes	29	[16]
2-Nitrophenol (NP)	2-chlorophenol (CP), Phenol (P)	Yes Yes	57	[16]
1,1,2-Trichloroethane (TCE-1)	1,1,2,2-Tetrachloroethane (TCE-2)	Yes	117	[7]
2-Chlorophenol (CP)	Phenol Nitrophenol (NP)	Yes Yes	148	[16]
1,1,2,2-Tetrachloroethane (TEC-2)	TCE-1	Yes	363	[7]
Chlorobenzene (CB)	Ethylbenzene (EB)	Slight	690	[16]
Tetrachloroethylene (TCE)	CB	No	759	[2]
Malathion (M)	Diazinon (D)	Yes	776	[9]
Diazinon (D)	Lindane (L) Lindane (L), Pentachlorophenol (PCP), Malathion (M)	No No	1380 1380	[9] [9]
Ethylbenzene (EB)	CB	No	1412	[9]
Lindane (L)	Pentachlorophenol (PCP)	No	5248	[9]
Phenanthrene (PT)	Dibutyl Phthalate (DP)	No	28,000	[2]
Pentachlorophenol (PCP)	L L,D,M	No No	44,668 44,668	[9] [9]
Dibutyl Phthalate (DP)	PT	No	1.58×10^5	[2]

Table 3.
Relationship between K_{OW} *and the competition effect in multicomponent sorption of toxic organic pollutants on wastewater solids.*

compound. For both methylene chloride and chloroform, the higher the concentration of 1,1-dichloroethylene, the more reduction in total uptake of the key component.

Based on the correlation given by Eq. (1), the degree of domination by either partitioning or surface adsorption depends on the values of K_{OW}. **Table 3** compares the K_{OW} values and the competition effect in several binary and multicomponent systems, for uptake of compounds by wastewater solids. It is clearly seen that the K_{OW} value can be qualitatively divided into three ranges: (1) $K_{OW} < 500$, there is a clear effect of the presence of a competing compound; (2) $500 < K_{OW} < 1000$, the competition effect is smaller, and depends on the specific compounds being studied; and (3) $K_{OW} > 1000$, when competitive surface adsorption can be ignored.

3. Analysis of sorption kinetics

The rate of partitioning of compounds into biomass depends on the rate of mass transfer of the compound into the cells, and this can be described as occurring in three steps: (1) mass transfer from the bulk water to the surface of the biomass cells; (2) transport through the cell walls; and (3) bonding with the active inter-cellular enzymes followed by biodegradation of the compound within the cell. In this section, these three steps will be quantified and experimental data on biodegradation rates will be used to analyze the extent of each step's contribution.

Cell membrane permeability has been shown to play an important role in the biodegradation process, when the membrane permeability is increased [17, 18]. The permeability of bilipid membranes can be estimated from the octanol–water partition coefficient and the molecular weight of the compound [13],

$$P = 0.003\frac{K_{OW}}{MW^{0.5}} \tag{2}$$

where K_{ow} is the octanol–water partition coefficient and MW is the molecular weight of the organic compound. The permeability coefficient can also be written as [19],

$$P = \frac{k_p D_{mean}}{\lambda} \tag{3}$$

Where k_P is the partition coefficient, D_{mean} is the average diffusion coefficient of the compound through the cell wall membrane, and λ is the thickness of the cell wall membrane. As discussed in the earlier section, the partition coefficient can be determined from the compound's K_{OW} value.

The dependance of a compound's diffusivity through the cell wall, D_{mem}, on molecular size, represented by molecular volume, is given by the following correlation:

$$\text{Log}D_{mem} = \text{Log}D_{mem}^{V=0} - m_V V \tag{4}$$

where $D_{mem}^{V=0}$ is the compound's diffusion coefficient for a theoretical molecule of infinitely small size, through the membrane. This correlation gives a slope of $m_V = -0.0013$ molecules/cm^3. From Eq. (3) we get the following result:

$$P^{V=0} = 10^{m_V V} P \tag{5}$$

where $P^{V\,=\,0}$ is the cell wall permeability for a molecule of infinitely small size and this limiting permeability values gives a good correlation with the octanol–water partition coefficient, K_{OW}, as shown in **Figure 4**.

According to Fick's first law of diffusion, the flux (J_b) of a compound from bulk liquid to the outside cell wall surface is given by the following equation:

$$J_b = k_b(C_b - C_{cm}) \tag{6}$$

where k_b is the mass transfer coefficient from the bulk to the cell surface, C_b and C_{cm} are concentrations of compound in the bulk liquid and at the cell membrane outside surface, respectively.

Flux (J_{cm}) through the cell membrane can be written as follows:

$$J_{cm} = P(C_{cm} - C_i) \tag{7}$$

where P is the compound's permeability through the cell membrane, C_{cm} is the compound concentration on the outside surface of the cell membrane and C_i is the

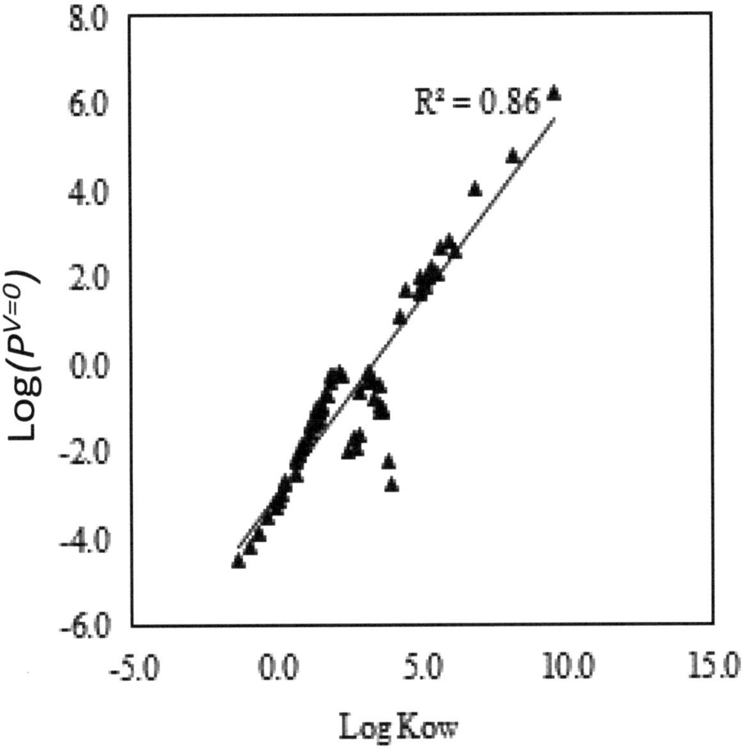

Figure 4.
Plot of Log($P^{V=0}$) versus LogK_{OW}. $P^{V=0}$ is the cell wall permeability for a molecule of infinitely small size.

intracellular compound concentration. Assuming that the Monod kinetics apply, the rate of substrate biodegradation inside the cell can be given by,

$$J_i = \frac{\mu_{max} C_i}{K_i + C_i}\left(\frac{V}{A}\right) \tag{8}$$

where J_i is the mass of substrate consumed per unit time per unit surface area of the cell, m_{max}, K_i are the maximum growth rate and half saturation constants, respectively, V is the volume of a cell and A is the surface area of the cell. For first order Monod kinetics, $K_x \gg C_i$ so Eq. (8) can be written as:

$$J_i = k_i C_i \left(\frac{V}{A}\right) \tag{9}$$

where $k_i = \mu_{max}/K_i$ is the first order biodegradation rate constant. Eqs. (6), (7), and (9) can be combined to give the following equation for the flux of the compound into the cell:

$$J_i = J_{cm} = J_b = J = \frac{C_b}{\left(\frac{1}{k_b} + \frac{1}{P} + \frac{A}{k_i V}\right)} \tag{10}$$

The flux of the compound into the cell can also be written as.

$$J = k_{\text{total}}(C_{\text{b}} - C_{\text{i}})\left(\frac{V}{A}\right) \quad (11)$$

where k_{total} is the overall mass transfer coefficient. The active intercellular enzymes will significantly reduce the intracellular concentration so that $C_{\text{i}} = 0$ Combining Eqs. (10) and (11) we get the following result:

$$\frac{1}{k_{\text{total}}} = \frac{1}{k_{\text{b}}}\left(\frac{V}{A}\right) + \frac{1}{P}\left(\frac{V}{A}\right) + \frac{1}{k_{\text{i}}} \quad (12)$$

The ratio (V/A) can be determined from the biomass concentration, X, in water, density of mixture of ρ_{i}, using the following equation, assuming all biomass cells are spherical, with an average diameter of d_i.

$$\frac{V}{A} = \frac{6X}{\rho_{\text{i}} d_{\text{i}}} = \alpha \quad (13)$$

Substituting Eq. (13) into Eq. (12) we get the following result (**Figure 5**).

$$\frac{1}{k_{\text{total}}} = \frac{1}{k_{\text{b}}}\alpha + \frac{1}{P}\alpha + \frac{1}{k_{\text{i}}} \quad (14)$$

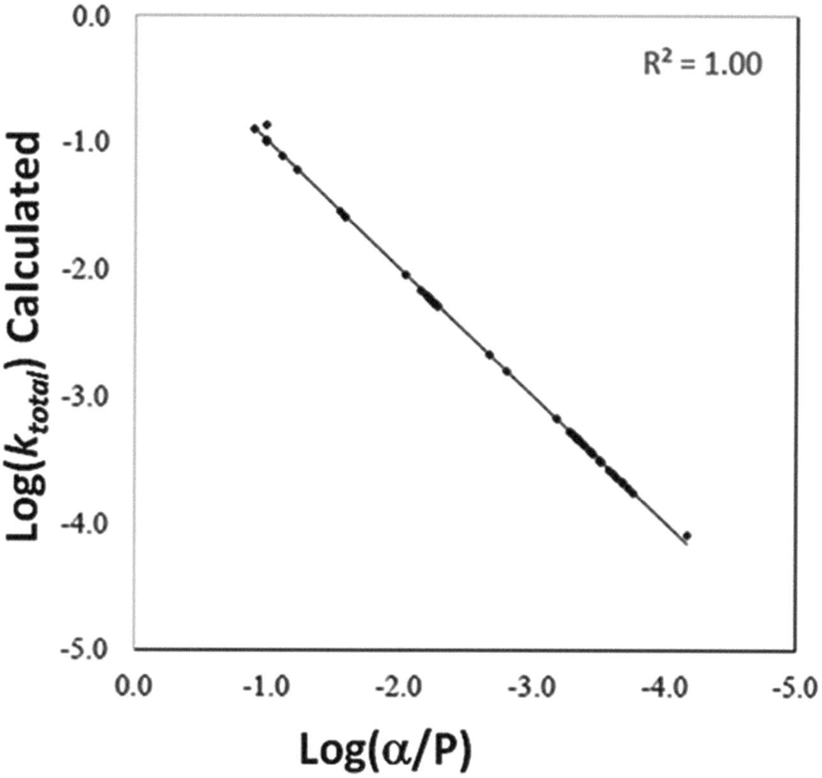

Figure 5.
Plot of calculated Log(k_{total}) vs. Log(α/P) showing that the overall biodegradation rate constant is actually diffusional resistance across the cell membrane for 50 randomly selected toxic organic compounds which were not used in the analysis.

Values for the overall mass transfer coefficient, k_{total}, were taken from an EPA report [20], the mass transfer coefficient from the bulk water to the outside cell surface, k_b, was estimated from the diffusivity of the compound in water assuming a boundary layer thickness of 4 micrometers [20], the cell permeability, P, was calculated using Eq. (2) assuming a biomass concentration, X, of 2.5 kg/m^3 [21], and a liquid density, ρ_i, as 1000 kg/m^3 and an average cell diameter, d_i of 1 micrometer [16, 22, 23].

Table 4 shows the calculated values of the bulk phase mass transfer coefficient, k_b, the cell wall permeability, P, calculated from Eq. (2), and the biodegradation kinetic

S. No	Compound name	$k_{total}\,(h^{-1})$	$\frac{k_b}{\alpha}\,(h^{-1})$	$\frac{P}{\alpha}\,(h^{-1})$	$k_i\,(h^{-1})$
1	Diisodecyl Pthalate	3.7659	765.8182	3.8827	1495.8965
2	Di-n-Octyl Phthalate	2.5381	706.9091	2.5780	2132.1180
3	Dibenzopyrene1,2,7,8	1.0573	981.8182	1.0727	799.1360
4	Bromacil	1.0087	1060.3636	1.0245	696.2582
5	Indeno (1,2,3-cd)-Pyrene	0.6596	1119.2727	0.6722	364.3080
6	Dimethyl Benz(A)ANT 7,12	0.5522	1197.8182	0.5634	284.7684
7	Dimethylbenz(A)Anthracene (7,12)	0.5522	981.8182	0.5628	300.4840
8	Methyl Cholanthrene 3	0.5394	1060.3636	0.5501	285.2126
9	PPCB's (Aroclors)	0.4850	1570.9091	0.4955	231.9249
10	Dieldrin	0.3411	922.9091	0.3483	167.9746
11	Benzo(B)Fluoranthene	0.3227	1099.6364	0.3298	153.2084
12	Benzo(K)Fluoranthene	0.3219	1099.6364	0.3289	152.7507
13	Toluene Diisocyanate (2,4)	0.3118	1217.4545	0.3187	145.4517
14	Tetra Chlorodibenzo-p-Dioxin (2,3,7,8)	0.2702	1138.9091	0.2762	124.9391
15	Methoxychlor	0.2546	883.6364	0.2601	120.5382
16	DDT	0.1823	981.8182	0.1864	82.3114
17	ChloroBenzilate	0.1790	1138.9091	0.1831	79.8019
18	Dichlorobenzophenone P,P	0.1608	1335.2727	0.1646	70.5525
19	Benzo(A)Pyrene	0.1517	1767.2727	0.1553	65.5411
20	Pentachlorobenzene	0.1499	1237.0909	0.1534	65.7798
21	Dimethyl Trisulfide	0.1329	1629.8182	0.1361	57.3433
22	Dibenzofurans	0.1183	1178.1818	0.1211	51.4321
23	DDE, p,p′ -	0.1177	1158.5455	0.1205	51.2012
24	Aldrin	0.1156	962.1818	0.1183	50.6902
25	Pentachloronitrobenzene	0.1132	1197.8182	0.1159	49.1063
26	Benzo(A)Anthracene	0.1097	1767.2727	0.1124	46.9333
27	Endrin	0.1088	922.9091	0.1113	47.6631
28	Hexachlorobenzene	0.0971	1158.5455	0.0994	41.9121
29	Tributyl Tin Acetate	0.0965	1138.9091	0.0988	41.6854
30	Tributyl Phosphate	0.0938	1021.0909	0.0960	40.6475

S. No	Compound name	$k_{total}(h^{-1})$	$\frac{k_b}{\alpha}(h^{-1})$	$\frac{P}{\alpha}(h^{-1})$	$k_i(h^{-1})$
31	Warfarin	0.0903	1060.3636	0.0925	39.0118
32	ChloroBenzylate	0.0891	903.2727	0.0912	38.7346
33	Trifluralin	0.0878	981.8182	0.0899	37.9977
34	Dodecane	0.0860	1158.5455	0.0881	36.9834
35	Triisobutylene	0.0839	1374.5455	0.0859	35.8866
36	Tetrachlorobenzene (1,2,3,4)	0.0771	1453.0909	0.0789	32.8341
37	Tetrachlorobenzene (1,2,3,5)	0.0771	1453.0909	0.0789	32.8330
38	Acifluorfen	0.0674	864.0000	0.0690	29.0357
39	Heptachlor	0.0672	1119.2727	0.0688	28.7305
40	ChloroazoBenzene	0.0659	1453.0909	0.0675	27.9799
41	Triisopropylamine	0.0624	1296.0000	0.0639	26.5454
42	Diisopropyl Benzene (Para)	0.0605	1413.8182	0.0619	25.6515
43	Methylene-Bis (2-Chloroaniline) 4,4′	0.0590	1138.9091	0.0604	25.1252
44	Tributyl Phosphorotrithioate SSS	0.0542	922.9091	0.0555	23.1364
45	Diphenylmethane	0.0539	1531.6364	0.0552	22.8083
46	TetrachloroPhenol (2,3,4,6)	0.0523	1394.1818	0.0535	22.1228
47	TetrachloroPhenol (2,3,5,6)	0.0523	1394.1818	0.0535	22.1223
48	Bisphenol(A)	0.0474	1119.2727	0.0486	20.1181
49	Benzophenone	0.0425	1296.0000	0.0435	17.9528
50	Tetrachlorobenzene (1,2,4,5)	0.0419	1728.0000	0.0429	17.6434
51	Trichlorobenzene 1,2,3	0.0396	1610.1818	0.0406	16.6779
52	Trichlorobenzene 1,3,5	0.0396	1610.1818	0.0406	16.6769
53	Diethylthiophosphatebenzo M Ethyl Pether	0.0357	1080.0000	0.0366	15.0790
54	BIS(1,1,2,2 - Tetrachloropropyl) Ether	0.0356	1040.7273	0.0364	15.0343
55	Biphenyl	0.0339	1610.1818	0.0347	14.2359
56	HexaFluoroacetone	0.0324	1394.1818	0.0331	13.6190
57	ChloroBenzophenone (PARA)	0.0322	1453.0909	0.0330	13.5492
58	Dimethylbenezidine 3,3	0.0311	1217.4545	0.0319	13.1076
59	Cymene, para	0.0308	1433.4545	0.0316	12.9555
60	Diethylbenzene P	0.0308	1590.5455	0.0316	12.9429
61	Methyl Napthalene (1-)	0.0301	1531.6364	0.0308	12.6524
62	Methyl Napthalene (2-)	0.0301	1531.6364	0.0308	12.6527
63	Chloronapthalene, 2-	0.0298	1728.0000	0.0305	12.5058
64	Silvex	0.0286	1138.9091	0.0293	12.0378
65	Chlorophazine	0.0257	1276.3636	0.0263	10.7930
66	Nitrobiphenyl, 4-	0.0256	1374.5455	0.0262	10.7449

S. No	Compound name	$k_{total}(h^{-1})$	$\frac{k_b}{\alpha}(h^{-1})$	$\frac{P}{\alpha}(h^{-1})$	$k_i(h^{-1})$
67	Pinene (alpha-)	0.0254	1433.4545	0.0260	10.6626
68	Acenapthene	0.0250	1512.0000	0.0256	10.4963
69	Phenylphenol P	0.0230	1335.2727	0.0235	9.6402
70	Diazinon	0.0229	962.1818	0.0235	9.6582
71	Ethylhexylacrylate 2-	0.0220	1197.8182	0.0225	9.2396
72	Ethyl Toluene, 4	0.0220	1531.6364	0.0225	9.2213
73	Tetralin	0.0218	1590.5455	0.0223	9.1447
74	Dichlorobenzonitrile,2,6-	0.0216	1492.3636	0.0221	9.0648
75	Hexachlorobutadiene	0.0214	1217.4545	0.0219	8.9730
76	Dipropylbutral	0.0212	1433.4545	0.0217	8.8953
77	Anthraquinone	0.0212	1492.3636	0.0217	8.8772
78	Phenylcyclohexanone 4	0.0205	1315.6364	0.0210	8.6025
79	Pentachloropheol	0.0204	1197.8182	0.0209	8.5610
80	Diisobutylene	0.0201	1433.4545	0.0206	8.4367
81	Dichlorophenol (2,4)	0.0198	1728.0000	0.0203	8.3269
82	Aminobiphenyl, 4-	0.0190	1492.3636	0.0195	7.9636
83	Octane	0.0188	1394.1818	0.0193	7.8944
84	Propyl (−n) Benzene	0.0188	1531.6364	0.0193	7.8850
85	Dimethyl Phthalate	0.0186	1237.0909	0.0190	8.0786
86	TrimethylPentane 2,2,4	0.0184	1472.7273	0.0188	7.6955
87	Diethyl (N,N) Aniline	0.0183	1158.5455	0.0187	7.6749
88	Endosulfan	0.0181	903.2727	0.0185	7.6105
89	Chlorotoluene-4	0.0179	1708.3636	0.0183	7.4858
90	Bromotoluene 4	0.0179	1669.0909	0.0183	7.4864
91	Dimethyl Amino Azobenzene, 4-	0.0176	1296.0000	0.0181	7.3979
92	Dazomet	0.0176	141.8182	0.0180	7.3761
93	DichloroBenzidine, 3,3′-	0.0175	1315.6364	0.0179	7.3261
94	Cumene	0.0173	1394.1818	0.0177	7.2422
95	Acetylaminofluorene, 2-	0.0170	1178.1818	0.0175	7.1539
96	Alpha Methyl Styrene	0.0168	2238.5455	0.0172	7.0159
97	Methylstyrene (−4)	0.0168	1728.0000	0.0172	7.0224
98	Hexachloroethane	0.0167	1335.2727	0.0171	6.9875
99	Trichloro-1,2,2,- Trifluoroethane, 1,1,2-	0.0167	1590.5455	0.0171	6.9805
100	Acenapthlyene	0.0162	1472.7273	0.0166	6.7824
101	Dimethyl Disulfide	0.0160	1983.2727	0.0164	6.6994
102	Carbendazim	0.0156	1276.3636	0.0160	6.5580
103	Methylene Diphenylamine (MDA)	0.0156	1256.7273	0.0160	6.5331

S. No	Compound name	$k_{total}(h^{-1})$	$\frac{k_b}{\alpha}(h^{-1})$	$\frac{P}{\alpha}(h^{-1})$	$k_i(h^{-1})$
104	Methylene Diphenyl Diisocyanate	0.0146	1217.4545	0.0150	6.1368
105	Toxaphene	0.0145	844.3636	0.0149	6.1114
106	Toluene	0.0144	1688.7273	0.0148	6.0352
107	Chlorambucil	0.0142	1178.1818	0.0146	5.9616
108	Dichlorophenol (2,6)	0.0138	1728.0000	0.0141	5.7780
108	Dichlorophenol	0.0138	1394.1818	0.0141	5.7826
110	Dichlorophenol 2,5	0.0138	1394.1818	0.0141	5.7841
111	Dichloroethane (1,1) ethylidenedichloride	0.0138	2061.8182	0.0141	5.9160
112	Nitrobenzene	0.0138	1688.7273	0.0141	5.9137
113	DichloroBenzene (1,4) (−p)	0.0138	1551.2727	0.0141	5.7758
114	Bromobenzene	0.0137	1826.1818	0.0140	5.7183
115	Xylene(−m)	0.0133	1531.6364	0.0136	5.5749
116	Anthracene	0.0132	1512.0000	0.0135	5.5204
117	Terpinenol, Alpha	0.0131	1453.0909	0.0134	5.4917
118	Heptane ISO	0.0129	1394.1818	0.0132	5.3957
119	Heptane (−n)	0.0129	1492.3636	0.0132	5.3896
120	Dimethoxy-(3,3′)-Benzidine	0.0128	1080.0000	0.0131	5.3488
121	Xylene(−p)	0.0127	1649.4545	0.0131	5.3348
122	EthylBenzene	0.0126	1527.7091	0.0129	5.2726
123	Nonanol, n	0.0120	1354.9091	0.0123	5.0072
124	Naphthol, alpha	0.0120	1492.3636	0.0122	5.0034
125	Nitro m Xylene, 2	0.0119	1570.9091	0.0121	4.9607
126	Parathion	0.0117	1138.9091	0.0120	4.9201
127	DiphenylHydrazine (1,2)	0.0115	1453.0909	0.0118	4.8052
128	DiChloroAniline 2,3	0.0114	1413.8182	0.0117	4.7824
129	DichloroAniline(2,3)	0.0114	1413.8182	0.0117	4.7832
130	Methyl Cyclohexane	0.0108	1669.0909	0.0111	4.5316
131	Xylene	0.0108	1826.1818	0.0111	4.5150
132	Xylene(−o)	0.0108	1963.6364	0.0111	4.5154
133	Diethylene Glycol Diethyl Ether	0.0106	1335.2727	0.0108	4.4195
134	Captan	0.0105	962.1818	0.0107	4.3904
135	Isodecanol	0.0104	1453.0909	0.0107	4.3643
136	Benzotrichloride	0.0104	1531.6364	0.0107	4.3649
137	Tetrafluoromethane	0.0102	1826.1818	0.0104	4.2524
138	2,4,5 Trichlorophenoxyacetic acid	0.0098	1315.6364	0.0101	4.1248
139	Dinitro-o-Cresol (4,6)	0.0098	1354.9091	0.0100	4.1089
140	Methyl Isocyanate	0.0098	2847.2727	0.0100	4.0764

S. No	Compound name	$k_{total}(h^{-1})$	$\frac{k_b}{\alpha}(h^{-1})$	$\frac{P}{\alpha}(h^{-1})$	$k_i(h^{-1})$
141	Naphthol, beta-	0.0097	1492.3636	0.0100	4.0679
142	Dichlorophenoxyacetic Acid (2,4)	0.0093	1276.3636	0.0095	3.9000
143	Chlordane	0.0092	864.0000	0.0094	3.8742
144	Carbon Tetrachloride	0.0090	1728.0000	0.0092	3.7650
145	Dichloropropylene 1,2,-(Cis)	0.0089	2160.0000	0.0091	3.7104
146	Hexane(−n)	0.0088	1531.6364	0.0090	3.6858
147	Xylidine Dimethylaniline	0.0088	1649.4545	0.0090	3.6854
148	Pentachloroethane	0.0087	1433.4545	0.0090	3.6556
149	Guthion	0.0087	942.5455	0.0089	3.6368
150	Trifluoroethane (1,1,1)	0.0086	2552.7273	0.0088	3.6057
151	Benzene	0.0084	1924.3636	0.0086	3.5278
152	Dichloropropane 1,2	0.0084	1708.3636	0.0086	3.5210
153	Chrysene	0.0084	1217.4545	0.0086	3.5101
154	Methyl 1-Pentene 2	0.0083	1767.2727	0.0085	3.4883
155	Propyl Ether Iso	0.0083	1826.1818	0.0085	3.4528
156	Styrene Oxide	0.0082	1747.6364	0.0084	3.4277
157	Ehtyl(2)Hexanol	0.0082	1433.4545	0.0084	3.4257
158	Octanol 3	0.0082	1433.4545	0.0084	3.4257
159	Octanol 2	0.0082	1433.4545	0.0084	3.4257
160	Octanol 4	0.0082	1433.4545	0.0084	3.4257
161	Octanol 1	0.0082	1433.4545	0.0084	3.4237
162	MethyleneDianiline 4,4	0.0081	1472.7273	0.0083	3.3975
163	Benzofuran 2,3	0.0081	1767.2727	0.0083	3.3710
164	ChloroNitrobenzene, p	0.0079	1845.8182	0.0081	3.3086
165	Chloronitrobenzene (−o)	0.0079	1845.8182	0.0081	3.3082
166	Chloroacetophnone, 2-	0.0078	1708.3636	0.0080	3.2837
167	Freon 11, Fluorotrichloromethane	0.0077	1963.6364	0.0079	3.2307
168	Dichloropropylene 1,2,- (Trans)	0.0077	2160.0000	0.0079	3.2068
169	ChloroBenzotriFluoride, P	0.0076	1688.7273	0.0078	3.1867
170	ButylIsoButyrate	0.0074	1728.0000	0.0076	3.1139
171	Trichlorofluoromethane	0.0074	1904.7273	0.0076	3.1019
172	Cyclohexane	0.0074	1786.9091	0.0076	3.0974
173	Dimethyl Benzylamine N,N	0.0074	1708.3636	0.0076	3.0938
174	Allyl Ether, diallyl ether	0.0073	1669.0909	0.0075	3.0718
175	Chlorophenol-4	0.0071	1904.7273	0.0073	2.9702
176	Nitrotoluene, o	0.0069	1708.3636	0.0071	2.9057
177	Nitrotoluene, p	0.0069	1688.7273	0.0071	2.9055
178	Nitrotoluene, m	0.0069	1610.1818	0.0071	2.9068

S. No	Compound name	$k_{total}(h^{-1})$	$\frac{k_b}{\alpha}(h^{-1})$	$\frac{P}{\alpha}(h^{-1})$	$k_i(h^{-1})$
179	ChloroButadiene, 1	0.0069	1963.6364	0.0071	2.9022
180	ChloroBenzyl Alcohol - (m)	0.0068	1865.4545	0.0070	2.8579
181	Triethylene Glycol Dimethyl Ether	0.0068	1315.6364	0.0070	2.8445
182	Freon 12, Dichlorodifluoromethane	0.0066	1669.0909	0.0068	2.7746
183	Ethylphenol,3-	0.0066	1845.8182	0.0068	2.7671
184	Tetraethyl Lead	0.0066	1256.7273	0.0067	2.7591
185	Methyl Benzyl Alcohol 4	0.0065	1688.7273	0.0067	2.7345
186	Carbaryl Sevin	0.0064	1394.1818	0.0065	2.6771
187	Triethylamine	0.0064	1551.2727	0.0065	2.6720
188	Phosphine	0.0063	3573.8182	0.0065	2.6302
189	Proporur (Baygon)	0.0062	1315.6364	0.0063	2.5915
190	Cumeme Hydroperoxide	0.0061	1492.3636	0.0063	2.5584
191	Paraldehyde	0.0061	1570.9091	0.0062	2.5471
192	Bromoform	0.0061	2022.5455	0.0062	2.5417
193	Benzyl Chloride	0.0061	1531.6364	0.0063	2.5349
194	Dibromo-4-HydroxyBenzonitrile,3,5	0.0060	981.8182	0.0062	2.5341
195	Dichloroethane (1,2)	0.0059	1944.0000	0.0060	2.5085
196	Naphthylamine, beta -	0.0058	1649.4545	0.0059	2.4275
197	Acetylmethylphthalate 4	0.0057	1099.6364	0.0058	2.3825
198	Naphthylamine, alpha -	0.0057	1649.4545	0.0058	2.3676
199	Diisopropylamine	0.0056	1531.6364	0.0057	2.3488
200	Dichloroethylene (1,2) Cis	0.0056	2218.9091	0.0057	2.3285
201	Diethylene Glycol Monobutyl Ether	0.0056	1374.5455	0.0057	2.3317
202	Butyl Carbitol	0.0056	1374.5455	0.0057	2.3310
203	TolueneSulfonyl Chloride	0.0055	1276.3636	0.0056	2.3096
204	Thiourea, 1-(o-Chlorophenyl)-	0.0054	1413.8182	0.0055	2.2656
205	Carbon Disulfide	0.0054	1963.6364	0.0055	2.2452
206	Freon 12 Dichlorodifluoromethane	0.0054	2061.8182	0.0055	2.2469
207	Benzal Chloride	0.0053	1865.4545	0.0055	2.2321
208	Chlorophenol-2	0.0053	1865.4545	0.0054	2.2282
209	Trichloroethylene	0.0053	1786.9091	0.0054	2.2093
210	TriPropylene Glycol	0.0052	1413.8182	0.0053	2.1848
211	Toluidine (−0)	0.0052	1786.9091	0.0053	2.1607
212	Toluidine m	0.0052	1806.5455	0.0053	2.1613
213	Dipropylamine	0.0051	2022.5455	0.0053	2.1505
214	Butyl Acrylate	0.0051	1512.0000	0.0052	2.1430
215	Toluic Acid (para-)	0.0051	1531.6364	0.0052	2.1441
216	Pentadiene 1,2	0.0050	2022.5455	0.0051	2.0718

S. No	Compound name	$k_{total}(h^{-1})$	$\frac{k_b}{\alpha}(h^{-1})$	$\frac{P}{\alpha}(h^{-1})$	$k_i(h^{-1})$
217	Dinitro Toluene 2,6	0.0049	1433.4545	0.0050	2.0460
218	Benzonitrile	0.0049	2002.9091	0.0050	2.0319
219	Methyl Parathion	0.0048	1158.5455	0.0049	2.0302
220	Quinoline	0.0048	1629.8182	0.0049	2.0086
221	Cyclopentadiene	0.0048	2140.3636	0.0049	1.9997
222	DinitroToluene (2,4)	0.0047	1394.1818	0.0048	1.9766
223	Cyclohexylamine	0.0047	2042.1818	0.0048	1.9466
224	Bromoxynil	0.0046	1021.0909	0.0047	1.9477
225	Chloropropylene-2	0.0046	2709.8182	0.0047	1.9276
226	Dichloropropene (1,3)	0.0046	1963.6364	0.0047	1.9226
227	Acrylonitrile	0.0045	2631.2727	0.0046	1.9706
228	Benzoyl Chloride	0.0045	2140.3636	0.0046	1.8869
229	Trichloroethane 1,1,1, Methyl Chloroform	0.0044	1728.0000	0.0045	1.8555
230	Trichloroethane 1,1,2	0.0044	1728.0000	0.0045	1.8590
231	Ethyl Morpholine, Ethyl Diethylene Oxime	0.0044	1728.0000	0.0045	1.8341
232	Methyl Chloride	0.0043	1276.3636	0.0045	1.8212
233	Nitrophenol,4-	0.0043	1885.0909	0.0044	1.8103
234	Methyl-Tertiary-Butyl Ether	0.0043	2061.8182	0.0044	1.7926
235	Ethylene Glycol MonoPhenyl Ether	0.0042	1649.4545	0.0043	1.7791
236	Bromodichloromethane	0.0042	2081.4545	0.0043	1.7656
237	Butadiene - (1,3)	0.0042	2120.7273	0.0043	1.7449
238	Butene	0.0041	2002.9091	0.0042	1.7362
239	Benzoic Acid	0.0041	1570.9091	0.0042	1.7335
240	Butane	0.0041	2199.2727	0.0042	1.7281
241	Furan	0.0041	2395.6364	0.0042	1.7190
242	Tetrachloroethene	0.0041	1610.1818	0.0042	1.7043
243	Tetrachloroethane (1,1,2,2)	0.0041	1551.2727	0.0042	1.7049
244	Dichloromonofluoromethane	0.0040	2258.1818	0.0041	1.6878
245	Benzidine	0.0039	2945.4545	0.0040	1.6619
246	Isobutylene	0.0039	2002.9091	0.0040	1.6351
247	Butyl Acetate (−n)	0.0039	1590.5455	0.0040	1.6334
248	Nitrophenol,2-	0.0039	1669.0909	0.0040	1.6288
249	Diethylene Glycol Dimethyl Ether	0.0039	1354.9091	0.0039	1.6204
250	Hexanol-1	0.0038	1472.7273	0.0039	1.6080
251	Dinitrophenol 2,4	0.0037	1786.9091	0.0038	1.5741
252	Amyl Acetate (−n)	0.0036	235.6364	0.0037	1.5311

S. No	Compound name	$k_{total}(h^{-1})$	$\frac{k_b}{\alpha}(h^{-1})$	$\frac{P}{\alpha}(h^{-1})$	$k_i(h^{-1})$
253	Dimethyl Sulfide	0.0036	2866.9091	0.0037	1.5114
254	Isophorone	0.0036	1335.2727	0.0037	1.5106
255	Methyl Iodide	0.0036	2042.1818	0.0036	1.4961
256	Benzyl Alcohol	0.0035	1767.2727	0.0036	1.4764
257	Adiponitrile	0.0035	1747.6364	0.0036	1.4764
258	DichloroBenzene (1,2) (−o)	0.0035	1551.2727	0.0036	1.4545
259	Tetraethyldithiopyrophosphate	0.0035	1080.0000	0.0035	1.4628
260	Trichloropropane (1,1,2)	0.0034	1826.1818	0.0035	1.4545
261	Vinyl Acetylene	0.0034	2768.7273	0.0035	1.4435
262	Ethyl Vinyl Ether	0.0034	1924.3636	0.0035	1.4363
263	Diethyl Ether	0.0034	1688.7273	0.0035	1.4284
264	Ethyl Ether	0.0034	1826.1818	0.0035	1.4281
265	Anisidine, o-	0.0033	1747.6364	0.0034	1.3925
266	Ethlene Dibromide	0.0033	2336.7273	0.0034	1.3862
267	Chloramben	0.0033	1669.0909	0.0033	1.3778
268	Hexanoic Acid	0.0033	1649.4545	0.0033	1.3734
269	Acetophenone	0.0032	1708.3636	0.0033	1.3597
270	Bis(2-Chloroethyl)Ether	0.0032	1472.7273	0.0033	1.3628
271	Furfural	0.0032	2042.1818	0.0033	1.3564
272	Cyclohexanol	0.0032	1629.8182	0.0033	1.3543
273	Ethylene Glycol Monobutyl Ether	0.0032	1610.1818	0.0032	1.3295
274	Butyl Cellosolve	0.0032	1590.5455	0.0032	1.3305
275	Dinitrobenzene M	0.0031	1492.3636	0.0031	1.2970
276	Ethyl Acrylate	0.0030	1688.7273	0.0031	1.2703
277	Chloral	0.0030	1904.7273	0.0031	1.2573
278	Ethylene Glycol Monobutyl Ether Acetate	0.0030	1335.2727	0.0030	1.2618
279	Dichloroethene 1,2 trans	0.0030	2336.7273	0.0030	1.2462
280	Benzaldehyde	0.0030	1786.9091	0.0030	1.2472
281	TolueneDiamine (2,6)	0.0029	1806.5455	0.0030	1.2216
282	TolueneDiamine (3,4)	0.0029	1806.5455	0.0030	1.2214
283	Toluene Diamine (2,4)	0.0029	1786.9091	0.0030	1.2225
284	Epoxybutane 1,2	0.0029	2022.5455	0.0029	1.2047
285	tetrahydrofuran	0.0029	2061.8182	0.0029	1.2050
286	Propylene	0.0028	2670.5455	0.0029	1.1914
287	Propene	0.0028	2336.7273	0.0029	1.1923
288	Chloroethane (Ethyl Chloride)	0.0028	2258.1818	0.0029	1.1920
289	Trichloropropane (1,2,2)	0.0028	1826.1818	0.0029	1.1964

S. No	Compound name	$k_{total}(h^{-1})$	$\frac{k_b}{\alpha}(h^{-1})$	$\frac{P}{\alpha}(h^{-1})$	$k_i(h^{-1})$
290	Propane	0.0028	2592.0000	0.0029	1.1847
291	Chlorophenol-3	0.0028	1845.8182	0.0028	1.1693
292	Toluidine p	0.0027	1845.8182	0.0028	1.1537
293	Methyl Isobutyl Ketone	0.0027	1531.6364	0.0028	1.1408
294	Trinitrotoluene (2,4,6)	0.0027	1256.7273	0.0027	2.0189
295	Propyl Acetate Iso	0.0027	1708.3636	0.0027	1.1239
296	Propyl (−n) Acetate	0.0027	1728.0000	0.0027	1.1234
297	Methyl Morpholine	0.0027	1767.2727	0.0027	1.1224
298	Chloro 2 Butene, 1 Trans	0.0026	1904.7273	0.0027	1.1134
299	Hexen-2-ONE 5	0.0026	1728.0000	0.0027	1.1083
300	Diethylene Glycol Monoethyl Ether	0.0026	1570.9091	0.0027	1.1035
301	Diethylene Glycol Monoethyl Ether	0.0025	1983.2727	0.0026	1.0507
302	Nitropropane 2	0.0025	1413.8182	0.0025	1.0446
303	HexaMethylene 1,6 Diisocyanate	0.0025	1276.3636	0.0025	1.0482
304	Diethylene Glycol Monoethyl Ether Acetate	0.0024	2454.5455	0.0025	1.0167
305	Diethyl Amine	0.0024	2297.4545	0.0025	1.0155
306	Methylene Chloride, Dichloromethane	0.0024	2002.9091	0.0024	0.9926
307	Dioxane (1,4)	0.0023	1158.5455	0.0024	0.9923
308	Ametryn	0.0023	1688.7273	0.0023	0.9700
309	Nitroaniline P	0.0023	216.0000	0.0023	0.9720
310	Phosgene (decomposes)	0.0023	3161.4545	0.0023	0.9536
311	Hydrogen Sulfide	0.0023	2906.1818	0.0023	0.9527
312	Methyl Mercaptan	0.0022	1590.5455	0.0022	0.9365
313Dibromoethane-1,2	0.0022	1845.8182	0.0022	0.9275	
314	Ethylene Glycol Dimethyl Ether	0.0022	1983.2727	0.0022	0.9208
315	Butyraldehyde ISO	0.0022	2395.6364	0.0022	0.9204
316	Dimethylethylamine	0.0022	2238.5455	0.0022	0.9202
317	Butyraldehyde	0.0022	2395.6364	0.0022	1.1617
318	Acrolein	0.0022	1826.1818	0.0022	0.9194
319	Ethylene Glycol MonoPropyl Ether	0.0021	2376.0000	0.0022	0.9018
320	Bromomethane	0.0021	2002.9091	0.0022	0.8841
321	Crotonaldehyde	0.0052	1786.9091	0.0053	2.1607
322	Vinyl Bromide	0.0021	2317.0909	0.0021	0.8773
323	Chloropropane-2	0.0020	1983.2727	0.0021	0.8609
324	Dimethyl Carbamoyl Chloride	0.0020	1963.6364	0.0021	0.8621
325	Quinone	0.0020	2101.0909	0.0020	0.8445

S. No	Compound name	$k_{total}(h^{-1})$	$\frac{k_b}{\alpha}(h^{-1})$	$\frac{P}{\alpha}(h^{-1})$	$k_i(h^{-1})$
326	Ethane	0.0019	3279.2727	0.0020	0.8140
327	Bromochloromethane	0.0019	1963.6364	0.0019	0.8134
328	Allyl Chloride	0.0019	2120.7273	0.0019	0.7973
329	Chloropropene-3	0.0019	2729.4545	0.0019	0.7952
330	Chloropropane-1	0.0019	2022.5455	0.0019	0.7908
331	Catechol	0.0018	1806.5455	0.0019	0.7854
332	Hexachloroxyclohexane (Gamma Isomer)	0.0018	1433.4545	0.0019	0.7904
333	Lindane Hexachlorocyclohexane	0.0018	1217.4545	0.0019	0.7945
334	Dichlorvos	0.0018	1433.4545	0.0019	0.7938
335	Vinyl Acetate	0.0018	1806.5455	0.0019	0.7799
336	EthylAcetate	0.0018	1904.7273	0.0019	0.7758
337	Methyl Isopropyl Ketone	0.0018	1806.5455	0.0018	0.7624
338	Diethylene Glycol Monomethyl Ether	0.0018	1688.7273	0.0018	0.7647
339	Propyl Amine Iso	0.0018	2061.8182	0.0018	0.7534
340	ButylAmine	0.0017	1885.0909	0.0018	0.7466
341	Dichloro-2-Butene, 1,2	0.0017	2022.5455	0.0018	0.7464
342	Dichloro-2-Butene, (1,4)	0.0017	1590.5455	0.0018	0.7465
343	Dichloro-2-Butene, 1,4	0.0017	1826.1818	0.0018	0.7502
344	Thiourea	0.0017	2709.8182	0.0018	0.7349
345	Phenylene Diamine (−o)	0.0017	1944.0000	0.0017	0.7272
346	Phenylene Diamine (−p)	0.0017	1944.0000	0.0017	0.7272
347	Phenylene Diamine (−m)	0.0017	1944.0000	0.0017	0.7251
348	Phthalimide	0.0017	1629.8182	0.0017	0.7302
349	Phthalic Acid	0.0017	1335.2727	0.0017	0.7152
350	Terephthalic Acid	0.0017	1394.1818	0.0017	0.7192
351	Caprolactam	0.0017	1767.2727	0.0017	0.7144
352	Cyclohexanone	0.0016	1688.7273	0.0017	0.7094
353	Cyanogen	0.0016	2690.1818	0.0017	0.7009
354	Acrylamide	0.0016	2081.4545	0.0017	0.6970
355	Resorcinol	0.0016	1708.3636	0.0017	0.6960
356	Butyric Acid	0.0015	1983.2727	0.0016	0.6611
357	Formaldehyde	0.0015	3888.0000	0.0015	0.6257
358	Dibromomethane	0.0015	1649.4545	0.0015	0.6528
359	Propionaldehyde	0.0015	2238.5455	0.0015	0.6351
360	Ethoxyethanol-2	0.0015	1885.0909	0.0015	0.6364
361	Dichloroethyl Ether	0.0015	1865.4545	0.0015	0.6432
362	Methacrylic Acid	0.0014	2061.8182	0.0015	0.6241

S. No	Compound name	$k_{total}(h^{-1})$	$\frac{k_b}{\alpha}(h^{-1})$	$\frac{P}{\alpha}(h^{-1})$	$k_i(h^{-1})$
363	Pyridine	0.0014	1492.3636	0.0015	0.6139
364	Nitroglycerin	0.0014	1531.6364	0.0015	0.6355
365	Methyl Ether Dimethyl Ether	0.0014	2925.8182	0.0014	0.6034
366	Chloroethylene	0.0014	3024.0000	0.0014	0.6030
367	Diethylhydrazine N,N	0.0014	2022.5455	0.0014	0.6022
368	Ethylene Glycol Monoethyl Ether Acetate	0.0014	1492.3636	0.0014	0.6138
369	Propylenimine 1,22 Methyl Aziridine	0.0014	2847.2727	0.0014	0.5962
370	Methyl Aziridine 2	0.0014	2847.2727	0.0014	0.5946
371	Dimethyl Hydrazine (1,1)	0.0014	2140.3636	0.0014	0.5858
372	Hydroquinone	0.0014	1767.2727	0.0014	0.5909
373	Aminophenol (−o)	0.0013	1688.7273	0.0014	0.5867
374	Aminophenol (−p)	0.0013	471.2727	0.0014	0.5874
375	Chloroprene	0.0013	1963.6364	0.0014	0.5756
376	Tamaron (Methamidiphos)	0.0013	1531.6364	0.0013	0.5692
377	Propanol	0.0012	2238.5455	0.0013	0.5397
378	Propiolactone b	0.0012	2238.5455	0.0013	0.5356
379	Tetranitromethane	0.0012	1374.5455	0.0013	0.5583
380	Methanol	0.0012	3220.3636	0.0012	0.8385
381	Methyl Ethyl Ketone, 2 Butanone	0.0012	1924.3636	0.0012	0.5352
382	Urethane	0.0012	2081.4545	0.0012	0.5202
383	Acetaldehyde	0.0012	2768.7273	0.0012	0.5096
384	Piperazine	0.0012	2042.1818	0.0012	0.5174
385	Chloropropionitrile,3-	0.0012	2454.5455	0.0012	0.5161
386	Chloroallyl Alcohol 2	0.0011	2415.2727	0.0011	0.4933
387	DimethylSulfoxide	0.0011	2179.6364	0.0011	0.4872
388	Neopentyl Glycol	0.0011	1806.5455	0.0011	0.4925
389	Butanedinitrile	0.0011	2317.0909	0.0011	0.4828
390	Dimethyl Sulfate	0.0011	1885.0909	0.0011	0.4803
391	Acrylic Acid	0.0011	2081.4545	0.0011	0.4682
392	Propanoic Acid	0.0010	2199.2727	0.0011	0.4606
393	Aminopyridine, 4-	0.0010	2120.7273	0.0011	0.4630
394	Propyn-1-Ol 2(Proparlgyl)	0.0010	2611.6364	0.0010	0.4468
395	Ethylene Glycol MonoMethyl Ether	0.0010	2199.2727	0.0010	0.4463
396	Ethylene Glycol Monoethyl Ether Cellosol	0.0010	1924.3636	0.0010	0.4505
397	Propylene Oxide	0.0010	1963.6364	0.0010	0.4366
398	Trichloropropane (1,2,3)	0.0010	1552.2727	0.0010	0.4409

S. No	Compound name	$k_{total}(h^{-1})$	$\frac{k_b}{\alpha}(h^{-1})$	$\frac{P}{\alpha}(h^{-1})$	$k_i(h^{-1})$
399	Trichloropropane 1,1,1	0.0010	1551.2727	0.0010	0.4342
400	Dibromo-3-Chloropropane,1,2	0.0010	1374.5455	0.0010	0.4542
401	Nitroso-N-Methylurea N	0.0010	2002.9091	0.0010	0.4248
402	Ethylene Glycol MonoMethyl Ether Acetate	0.0010	1570.9091	0.0010	0.4406
403	Ethylamine	0.0009	3298.9091	0.0010	0.4190
404	Aziridine ethylene imine	0.0009	3102.5455	0.0010	0.4144
405	Acetaldol	0.0009	2120.7273	0.0010	0.4232
406	Allyl Alcohol	0.0009	2238.5455	0.0010	0.4118
407	Dipropylene Glycol	0.0009	1512.0000	0.0010	0.4185
408	Tetraethylene Pentamine	0.0009	1276.3636	0.0009	0.4169
409	Xylenol(3,4)	0.0009	1629.8182	0.0009	0.4092
410	Methyl Formate	0.0009	2493.8182	0.0009	0.3950
411	Nitromethane	0.0009	2749.0909	0.0009	0.3849
412	Vinyl Chloride	0.0009	2415.2727	0.0009	0.3870
413	Dimethyl Amine	0.0008	3279.2727	0.0009	0.3783
414	DimethylAcetamide	0.0008	2415.2727	0.0009	0.3789
415	DimethylSulfoxide	0.0008	2592.0000	0.0009	0.3753
416	Epichlorohydrin	0.0008	1924.3636	0.0009	0.3753
417	Glutaric Acid	0.0008	1570.9091	0.0008	0.3752
418	HexamethylPhosphoramide	0.0008	1354.9091	0.0008	0.3877
419	Propylene Chlorohydrin	0.0008	2061.8182	0.0008	0.3726
420	Dimethyl Formamide	0.0008	2022.5455	0.0008	0.3564
421	Butylene Glycol - (1,3)	0.0008	2002.9091	0.0008	0.3604
422	Methyl Acetate	0.0007	1963.6364	0.0008	0.3504
423	Diazomethane	0.0007	3436.3636	0.0007	0.3248
424	Adenine	0.0007	1708.3636	0.0007	0.3469
425	Chloromethyl Methyl Ether	0.0007	2631.2727	0.0007	0.3201
426	Chloroacetaldehyde	0.0007	2258.1818	0.0007	0.3168
427	Butanol(S)	0.0007	2199.2727	0.0007	0.2783
428	Butanol-1	0.0007	1826.1818	0.0007	0.2778
429	Butanol Iso	0.0007	1826.1818	0.0007	0.2781
430	DimethylSulfone	0.0006	1944.0000	0.0007	0.3156
431	Ethlene Diamine	0.0006	2768.7273	0.0007	0.2986
432	Dichloro Propanol 2,23	0.0006	1924.3636	0.0007	0.3229
433	Carbonyl Sulfide	0.0006	2552.7273	0.0007	0.2936
434	Diethyl Sulfate	0.0006	1590.5455	0.0006	0.3067
435	Methomyl	0.0006	1413.8182	0.0006	0.3234

S. No	Compound name	$k_{\text{total}}(\text{h}^{-1})$	$\frac{k_b}{\alpha}(\text{h}^{-1})$	$\frac{P}{\alpha}(\text{h}^{-1})$	$k_i(\text{h}^{-1})$
436	Acetyl Chloride	0.0006	2258.1818	0.0006	0.3021
437	Acetonitrile	0.0006	3259.6364	0.0006	0.2786
438	Bis(Chloromethyl)Ether	0.0006	1845.8182	0.0006	0.2984
439	Glycidol	0.0006	2395.6364	0.0006	0.2774
440	Succinic Acid	0.0006	1669.0909	0.0006	0.3003
441	Sodium Formate	0.0006	3200.7273	0.0006	0.2680
442	Nitrosomorpholine	0.0006	1963.6364	0.0006	0.2879
443	Maleic Acid	0.0005	2258.1818	0.0006	0.2779
444	Fumaric Acid	0.0005	1688.7273	0.0006	0.2895
445	Ethyl Carbamate	0.0005	2474.1818	0.0006	0.2834
446	Urea	0.0005	2690.1818	0.0005	0.2649
447	Nitrosodimetylamine N	0.0005	2434.9091	0.0005	0.2724
448	Methyl Hydrazine	0.0005	2729.4545	0.0005	0.2567
449	Propylene Glycol	0.0005	2002.9091	0.0005	0.2648
450	Dichloro 2-Propanol 1,3	0.0005	1924.3636	0.0005	0.2671

Table 4.
Values of mass transfer coefficients for overall transport of toxic organic compounds ($1/k_{total}$), obtained from Ref. [19], bulk diffusion from bulk water to outside cell surface (α/k_b), across the cell membrane (α/P) and biodegradation rate constant ($1/k_i$).

constant, k_i, determined from Eq. (12), using the values of k_{total} from EPA's report [20]. The calculated values clearly show that the cell wall permeability is the controlling resistance. It also shows that the measured values of the biodegradation kinetic constant, as given in EPA's Report [20], are mainly resistance of the compound's diffusion across the cell wall membrane.

To check this surprising finding, 50 organic compounds, listed in **Table 5**, were randomly selected from EPA's report [20]. These compounds were not used to correlate the overall degradation rate with the cell wall permeability. **Figure 4** shows the plot of the reported Log(k_{total}) values for these 50 compounds [20] versus the mass transfer resistance for pollutant transport across the cell wall, (α/P). This resulted in an excellent correlation, indicating that cell wall permeability is the rate controlling step for biodegradation, as well as the rate controlling step for sorption, which precedes biodegradation.

The kinetics of sorption is the rate of mass transfer to the cell wall followed by permeation of the compound across the cell membrane. Between these two transport steps, permeation across the cell wall is the rate controlling step. This can be written as follows

$$\frac{1}{k_{\text{sorption}}} \approx \alpha\left(\frac{1}{P}\right) \quad (15)$$

Table 4 shows that permeation of the compound through the cell membrane is the rate controlling step. The kinetic biodegradation rate constants, as presented in the

S. No	Compound name
1	1,3-Dichloro 2-Propanol
2	Glyphosate
3	Methyl Amine
4	Formamide
5	Phthalic Anhydride
6	Chloroacetic Acid
7	Ethylene Thiourea
8	Acetyl-2-thiourea, 1-
9	Diethanolamine
10	Iso-Propanol
11	Monomethyl Formamide
12	Ethanolamine (mono-)
13	Acetamide
14	*N*,*N*-Dimethylaniline
15	Ethylene Glycol
16	Formic Acid
17	Chlorohydrin
18	Morpholine
19	Hydroxyacetic Acid
20	Glyoxal
21	Oxalic Acid
22	Hydrazine
23	Glycerin (Glycerol)
24	Dibromochloromethane
25	Pentaerythritol
26	Nabam
27	Triethanolamine
28	1,3-Propane Sultone
29	Styrene
30	2,4,6-Trichlorophenol
31	Bis(2-Ethylhexyl) Phthalate
32	p-Chloroaniline
33	2-Chloroaniline
34	3-Chloroaniline
35	Ethylene Oxide
36	Aldicarb
37	Vinylidene Chloride
38	Diisopropyl Ketone

S. No	Compound name
39	Ethanol
40	Dichloroethene (1,1) vinylidene chloride
41	Maleic Anhydride
42	Napthalene
43	DibutylPhthalate
44	Acetal
45	Toluic Aldehyde
46	Acetic Acid
47	Phosphine
48	Proporur (Baygon)
49	Cumeme Hydroperoxide
50	Paraldehyde

Table 5.
List of 50 toxic organic compounds selected from the EPA report [19], which were used to test the finding that the biodegradation kinetics, as reported in Ref [19] is actually a measurement of diffusional resistance across the cell membrane.

EPA report [20], also given in **Table 4**, are actually sorption rate constants, since the cell wall permeability is the dominant rate controlling step.

4. Conclusions

In this chapter, sorption in wastewater solids has been presented both from a thermodynamic viewpoint, with surface adsorption and sorption into the cells represented by an equilibrium isotherm. Experimental data on the equilibrium concentrations of several compounds was analyzed using a multicomponent model. The experimental data shows that for compounds with low K_{OW} vales (K_{OW} < 500), competitive surface adsorption occurred, indicating that the multiple compounds were competing for the limited number of surface adsorption sites on the biomass. For compounds with K_{OW} vales in the 500–1000 range, surface adsorption dominated and there was less competition between the compounds for adsorption. When the K_{OW} value exceeded 1000, there was insignificant competition between the compounds for surface adsorption and sorption into the cell was dominant. The reason for diminishing competition between the compounds in a multi-component system was due to increasing partitioning of the compounds within the cell, with increases in K_{OW} above 500. Partitioning of compounds into the cell dominates when the K_{OW} value exceeds 1000.

Sorption of compounds into biomass has also been presented from a kinetic point of view by detailing the compound's transport across the cell membrane. The rate of a compounds transfer from bulk water into the biomass cells was sub-divided into three steps: (1) transport from the bulk water to the outside of the cell walls; (2) permeation through the cell membrane walls; and (3) bonding with active intercellular enzymes followed by biodegradation. Each step in the sorption process was further modeled and values of the mass transfer coefficients for each step were calculated using EPA's data on the overall biodegradation rate for over 500 compounds. This analysis showed that the rate controlling step in sorption was permeation of the compound through the cell walls, and the rate of mass transfer to the cell wall and biodegradation within the cell were not major contributors.

Analysis of sorption kinetics clearly shows that experimental biodegradation rates in the literature, as summarized in EPA report [20] are actual rates of permeation through the cell wall and the use of Monod kinetics, which has been presented as an enzymatic process of biodegradation, is actually a passive diffusive process through the walls of the microbial cells.

Author details

Rakesh Govind* and Ankurman Shrestha
University of Cincinnati, Cincinnati, OH, USA

*Address all correspondence to: govindr@ucmail.uc.edu

References

[1] Al-Ghouti MA, Al Absi RS. Mechanistic understanding of the adsorption and thermodynamic aspects of cationic methylene blue dye onto cellulosic olive stones from wastewater. Scientific Reports. 2020;**10**:15928

[2] Dobbs RA, Wang L, Govind R. Sorption of toxic organic compounds on wastewater solids: Correlation with fundamental properties. Environmental Science and Technology. 1989;**23**(9): 1092

[3] Wang L, Govind R, Dobbs RA. Sorption of toxic organic compounds on wastewater solids: Mechanism and modeling. Environmental Science and Technology. 1993;**27**:152

[4] Sugiura K, Sato S, Goto M. The adsorption-diffusion mechanism of BHC-residues. Chemosphere. 1975;**4**:189

[5] Wedemeyer G. Uptake of 2,4-Dichlorophenoxyacetic acid by *Pseudomonas fluorescens*. Applied and Environmental Microbiology. 1966;**14**: 486

[6] Amy GL, Bryant CW, Alleman BC, Barkley WA. Biosorption of organic halide in a Kraft mill generated lagoon. Journal—Water Pollution Control Federation. 1988;**60**(8):1445

[7] Tsezos M, Seto W. The adsorption of chlroethanes by microbial biomass. Water Research. 1986;**20**(7):851

[8] Bell JP. Biosorption of hazardous organic pollutants. [PhD thesis]. Canada: McMaster University; 1987

[9] Bell JP, Tsezos M. Removal of hazardous organic pollutants by adsorption on microbial biomass. Water Science and Technology. 1987;**19**:409

[10] Bell JP, Tsezos M. Removal of hazardous organic pollutants by biomass adsorption. Journal—Water Pollution Control Federation. 1987;**59**(4):91

[11] Bell JP, Tsezos M. The selectivity of biosorption of hazardous organics by microbial biomass. Water Research. 1988;**22**(10):1245

[12] Torres E. Biosorption: A review of the latest advances. PRO. 2020;**8**:1584

[13] Chiou CT. Partition coefficients of organic compounds in lipid-water system and correlations with fish bioconcentration factors. Environmental Science and Technology. 1985;**19**(1):57

[14] Chiou CT, Peters LJ. Freed VH. A physical concept of soil-water equilibria for nonionic organic compounds. Science. 1979;**206**(16):831

[15] Chiou CT, Porter PE, Schmedding DW. Partition equilibria of nonionic compounds between soil organic matter and water. Environmental Science and Technology. 1983;**17**(4):227

[16] Ouzounov N et al. Orientation correlates with cell diameter in *Escherichia coli*. Biophysical Journal. 2016;**111**:1035-1043

[17] Wang L et al. Biosorption and degradation of decabromodiphenyl ether by *Brevibacillus brevis* and the influence of decabromodiphenyl ether on cellular metabolic responses. Environmental Science and Pollution Research. 2016;**23**: 5166-5178

[18] Liao L et al. Biosorption and biodegradation of pyrene by *Brevibacillus brevis* and cellular responses

to pyrene treatment. Ecotoxicology and Environmental Safety. 2015;**115**:166-173

[19] Bressler DC, Gray MR. Transport and reaction processes in bioremediation of organic contaminants. 1. Review of bacterial degradation and transport and reaction processes in bioremediation of organic contaminants. International Journal of Chemical Reactor Engineering. 2003;**1**:1–16

[20] Air emission models for waste and wastewater. EPA-453/R-94-080A. 1994. pp. C5–C30

[21] Wioland HC, Woodhouse FG, Dunkel J, Kessler JO, Goldstein RE. Confinement stabilizes a bacterial suspension into a spiral vortex. Physical Review Letters. 2013;**110**:1-5

[22] Govind R, Gao C, Lai L, Tabak HH. Continuous automated and simultaneous measurement of oxygen uptake and carbon dioxide evolution in biological systems. Water Environment Research. 1996;**69**:73-80

[23] Selvakumar A, Hsieh HN. Competitive adsorption of organic compounds by microbial biomass. Journal of Environmental Science and Health. 1988;**A23**(8):729

Chapter 2

Coconut Shell Charcoal Adsorption to Remove Methyl Orange in Aqueous Solutions

Isabel Cristina Páez-Pumar Romer,
Isabella Victoria Plazola Santana,
Rosa María Rodríguez Bengoechea
and Miguel Manuel Pérez Hernández

Abstract

Activated charcoal was prepared and characterized from residues of coconut peel (CACC) to remove by adsorption the Methyl Orange (AM) dye in aqueous solution. The charcoal was activated with phosphoric acid. The morphology and structure of the pores of the carbon obtained were analyzed by Scanning Electron Microscopy (SEM) and a surface analyzer. The adsorption data were evaluated by the BET, Langmuir and Freundlich isotherms, finding the Langmuir type I model. The surface area of the activated carbon was 526 m^2/g with a pore volume of 0.234 cm^3/g and an average pore diameter of 1.78 nm, according to BET, which indicates the presence of micropores. The calculated thermodynamic parameters showed that the adsorption of the AM dye in CACC is a spontaneous process at room temperature and that physisorption and chemisorption are probably involved. The adsorption tests were followed by UV–visible spectrophotometry. The effects of the adsorbate concentration (AM) and the heat treatment (450–500°C) with an air atmosphere were investigated, keeping constant the stirring time and the H_3PO_4/sample weight ratio. The results obtained indicate that the activated carbon obtained could be used as an alternative low-cost adsorbent in the removal of AM from effluents in aqueous solution.

Keywords: waste, colorant, Langmuir, spontaneous, chemisorption, spectrophotometer

1. Introduction

With economic and technological development, water pollution is a common problem around the world, particularly in the textile, printing paper, pharmaceutical, food manufacturing industry, and in research laboratories [1–3]. Activated charcoal

(AC) shows great capacity as adsorbent in water purification or industrial effluent treatment due to its high pore volume, large specific surface area, high degree of surface reactivity, and effective adsorption quality [4–6]. This adsorption method allows the removal of up to 90% of pollutants; however, the process efficiency will depend on the physicochemical properties of the adsorbent and adsorbate [7]. Additionally, the high-cost and commercially available nonrenewable source of AC limits its use as an adsorbent in developing countries [8]. In recent years, the production of AC from cheap and renewable precursors (recyclable material), such as walnut shells, fruit seeds, pineapple, bagasse, bamboo, rice husk, cotton stems, eggshell, has been studied [9–12]. Coconut shell is a potential precursor for AC production due to its excellent natural structure and low ash content.

Converting coconut shells to activated charcoal would add value to these agricultural products, help reduce the cost of waste disposal, and provide a potentially inexpensive alternative to existing commercial charcoals [13]. The charcoal obtained from this recycling material could function as a good colorant remover in the water purification process, through adsorption processes, as proposed in this work. It is chosen to work with methyl orange (MO) as adsorbent because it is one of the anionic dyes most used in textile industries, [14, 15]; thus, for the purposes of this research, it is considered representative. The adsorption process consists mainly of two stages: first, the passage of the adsorbate through the porous network (diffusion) and the second, the interactions between the adsorbate and the adsorbent. This last step constitutes an important factor considering that some pharmaceutical compounds and organic dyes have diffusive limitations due to their large molecular size [16, 17].

For the preparation of activated charcoal, there are two methods that are mainly used: physical activation (heat treatment, at temperatures that can vary between 400 and 650° C) and chemical activation. In chemical activation, prior heat treatment, the precursor is reacted with an activating chemical agent. Acids, alkalis, and salts in solid state or solution are often used as chemical activating agents. These activating agents promote dehydration and then structural rearrangement at relatively low temperatures [18, 19].

Although porosity is the main characteristic related to the adsorption capacity of activated charcoal (this property will not be determined from a quantitative point of view in this work), the surface chemical composition also plays an important role in said adsorption when specific physical and/or chemical interactions are considered because it determines the moisture content, the catalytic properties, their acidic or basic character, and adsorption of polar species [20, 21].

The main objective of this research is to synthesize and evaluate the charcoal adsorptive capacity that is obtained from the recycling material and verify its adsorption capacity through the removal of methyl orange dye (MO) in aqueous solutions. The operation effects such as the initial concentration of MO were studied, a single type of activating agent was maintained (phosphoric acid), the same average particle size of charcoal was used, the temperature and activation time were a constant [19], the same charcoal/activator ratio and the same initial pH were preserved.

2. Material and methods

2.1 Raw material treatment

The coconut, *Cocos nucifera L.*, is a perennial tropical crop, which involves more than 4000 agricultural producers in Venezuela, and in recent years, the coconut production has ranged between 130,000 and 178,000 tons [22]. From the coconut that is formed by the shell or husk (endocarp), water, or copra (flesh), the endocarp represents about 30% weight [22].

The coconut shells constitute a recycling material, which are collected from different places located in Chacao municipality, Caracas, Venezuela. Initially, these shells were subjected to a conditioning process, in order to remove the pulp remains that were adhered to the shell and separate the endocarp from the rest of the coconut. Afterward, they were crushed to reduce their size to smaller particles suitable for sample carbonization. Later, they were sieved up to a size of 2–4 mm to obtain uniform samples. Finally, they were stored in a desiccator.

2.2 Preparation and chemical activation of charcoal samples

The samples were prepared by a chemical activation in two steps. In the first step, after the coconut shells were sieved, they were carbonized in a ceramic crucible in air atmosphere, at atmospheric pressure, up to temperatures of 450°C and 500°C (heat treatment), respectively. According to previous research [23], initially, the samples were placed in the horizontal muffle (brand Felisa FE-340) at 300°C. Fifteen minutes later, the temperature was increased to 50°C/15 minutes, and they were kept for 1 hour at the chosen temperatures. Then, the carbonized samples were crushed and sieved again, in order to achieve particle sizes between 150 and 250 μm. In the second step, after carbonization (heat treatment), the samples were impregnated with a solution that was prepared with distilled water at 80% v/v de H_3PO_4 (reagent grade ≥ 85% weight, Honeywell Riedel-de Haën AG), in a precipitation flask, in the ratio 1:2, g charcoal/m phosphoric acid. Then, the samples were dried in an oven at 105°C for a period of 12 hours [24]. Afterward, the sample was activated, which was introduced into the same muffle at a heating rate of 5°C/ up to the final temperature of 600°C in air atmosphere, for a period of 2 hours. The product was cooled to room temperature and washed according to the methodology described by De la Hoz, et al. [25], to remove the phosphoric acid residues that are present in it. The washing was carried out with distilled water at a 0.1 M up to pH 7. To conclude, the samples were dried in the oven at 100°C for 24 hours to remove traces of water from the charcoal.

2.3 Studied adsorbate

Methyl Orange (MO), also called Helianthine or Acid Orange 52, is an anodic dye, which is soluble in water and is used in the textile industry as an acid–base indicator. Its commercial compound name is 4-dimethylaminoazobenzene sulfonic acid sodium salt. Its molecular formula is $C_{14}H_{14}N_3NaO_3S$, and its molecular weight is 327.34 g/mol. **Figure 1**.

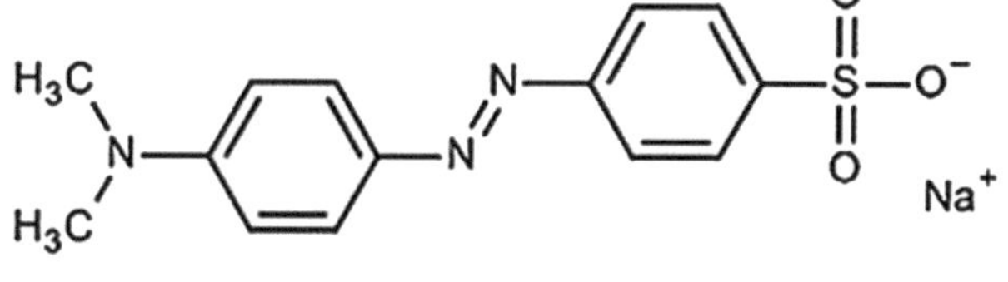

Figure 1.
Developed structure of OM, source: Academic (2010) [26].

The MO is in a solid state as a yellow or orange powder or crystals, it has not odor and is soluble in water. As MO is composed of aromatic rings, it is carried out by interactions between the π bonds of the aromatic ring and those that are found in the charcoal surface [27]. In the basic form of MO, the hydrogen ion is lost from the bridge –N=N- between the rings, and the electrons used to bind the hydrogen neutralize the positive charge on the terminal nitrogen, so that it is no longer able to form the bond π [28] (**Figure 2**). The color of MO solutions is yellow in alkaline medium.

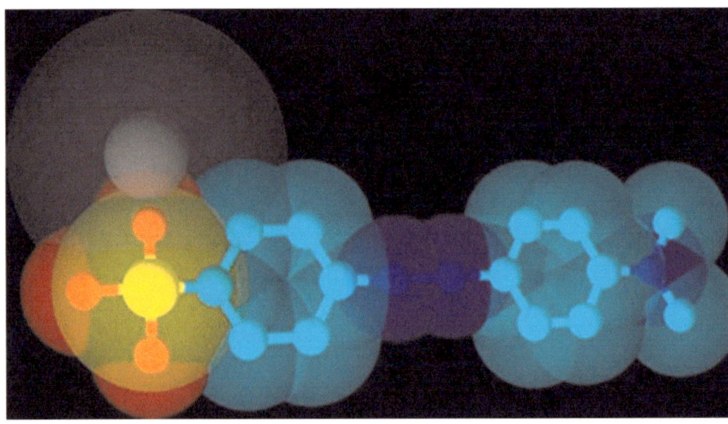

Figure 2.
*View of the methyl orange molecule with the presence of Van der Waals forces. [**HHH**], edited with ACD/3D viewer 2019.2.0 (file version D05E41, build 108,653, Apr 25, 2019).*

This azo-derived chemical compound with weak acid characteristics presents health risks in an aqueous medium, which is why several water treatments are practiced for its removal, such as degradation under irradiation or electrocoagulation [29].

D. A. Kron et al. demonstrated that Methyl Red (sodium salt) has dimensions of 1.61 nm length and 0.88 nm width [30]. Because the MO has a similar structure to methyl red, it is reasonable to expect that MO dimensions to be similar to methyl red. Methyl Orange (MO) is one of the most widely used anionic dyes in these industries, the presence of azo group (-N=N-) in its molecule, its low biodegradability, and its high solubility in water make it a serious threat for the environment [14, 15].

3. Synthetic solution preparation of methyl orange dye

Residual aqueous solutions of dye were simulated with different concentrations of MO (400, 200, 80 y 40 mg/L) [31]. The solutions were prepared with distilled water and analytical reagent grade MO (Merck).

3.1 Characterization of coconut shell charcoal

The porous structure of the coconut shell charcoal obtained was characterized by the nitrogen adsorption–desorption technique using the automatic instrument TriStar ® 3000 V2.0 from Micromeritics I. Corp., USA., at liquid nitrogen temperature (77 K). The specific surface area (SBET) of activated charcoal was calculated by using the Brunauer–Emmett–Teller (BET) equation, assuming that the area of nitrogen molecule is 0.162 nm^2. The total pore volume was estimated as the liquid volume of adsorbate adsorbed (N_2), at a relative pressure (P/Po) of 0.99 (**Table 1**).

Area	
BET surface area	526.47 m^2/g
Micropore area, t-Plot	463.75 m^2/g
External surface area, t-Plot	62.72 m^2/g
Volume	
Total pore Volume (a)	0.2341 cm^3/g
Adsorbed volume	120.94 cm^3/g STP
Micropore volume, t-Plot.	0.1762 cm^3/g
Mesopore volume, BJH (b)	0.03641 cm^3/g
Pore Volume	
Average pore diameter (4 V/A, BET)	1.7784 nm
Average pore diameter (4 V/A, BJH)	7.7859 nm

Source: Materials characterization laboratory report, INTEVEP.

Table 1.
Texture properties that were obtained through the adsorption/desorption studies of N_2. Parameters of the charcoal obtained. (a) Adsorption, total pore volume of less than 212.7 nm in diameter of the charcoal obtained at P/Po 0.99; (b) adsorption, cumulative pore volume of pores between 1.7 and 300 in diameter.

The micropore volume and the microporous surface area were determined by the t-Plot method [32]. The external surface area (which is the surface covered by mesopores and macropores) was calculated by the difference of the BET surface area, and the microporous surface was calculated from the t-Plot graph. The t-Plot method that was designed by De Boer et al. and Sing and referenced by Coasne et al. [33], is a well-known technique that allows determining the micro- and/or mesoporous volumes and the specific surface area of a sample, in comparison with a reference adsorption isotherm of a nonporous material that has the same surface chemistry. In order to observe the morphology of the activated charcoal that is obtained from the coconut shell, Scanning Electron Microscopy was used. For the sample preparation, the cathodic sputtering technique was used. The process was carried out under a pressure of 0.1 mbar and 20 mA for 20 minutes. (Scanning Electron Microscope, Brand JEOL®, Model JSM-6390).

4. Adsorption studies

The adsorption experiments were conducted in batch or discontinuous mode, mixing 25 mL of a dye solution of known concentration of 0.1 of charcoal, using the charcoals prepared at 450°C and 500°C [34]. The effect of the initial concentration of MO (400 and 200 mg/L), contact time (2 hours) [34], and temperature (25°C) was studied on the MO adsorption with stirring speed of 200 rpm and initial pH indicated in **Table 2**.

It is known that the pH of the solution is important in the adsorption of MO on Activated Charcoal [35–37]. The highest adsorption capacity is obtained toward acid pH, between pH 3 and pH 5. Different authors attribute this effect to the variation in the properties of the adsorbent surface and also to the ionization degree [38].

pH	5.87	5.85	5.77	5.76
C(MO), mg/L	40	80	200	400

Table 2.
Initial pH value for different MO concentrations, in mg/L.

After the adsorption, the samples were filtered by gravity in a precipitation flask to obtain a charcoal-free solution. The resulting liquid was centrifuged for 15 minutes at a stirring speed of 2400 rpm using a tabletop centrifuge (Gemmy Industrial Corp. ® Modelo PLC-05, Taiwan), in order to separate the residual charcoal particles. Finally, the supernatant that was contained in the centrifuge tube was sucked out. The supernatant was analyzed by a calorimeter method using a UV–visible spectrophotometer, brand Fisher ® 4001/4, Thermo Scientific, Genesys 20, U.S.A, at a wavelength of 520 nm [39–41].

MO concentrations in aqueous solution were determined by using the data of the calibration curve, concentration based on absorbance, by means of Eq. (1):

$$C = k.Abs + b \tag{1}$$

Where C is the concentration of methyl orange, mg/L, Abs is the absorbance, and k, b are the constants for the adjustment. Since the spectrophotometer was calibrated with distilled water, the b term is equal to zero. The MO concentration was established by measuring the absorbance and substituting it in the equation of the calibration curve. With the calibration curve obtained and the absorbance indicated by the equipment, the final concentration of solution was calculated, and later, based on the results obtained, the optimum carbonization temperature was determined. The tests were carried out in triplicate.

The MO adsorption capacity at equilibrium, qe (mg/g), and the removal efficiency Q (%) were obtained according to the Eqs. (2) and (3), respectively:

$$qe = \frac{(Co - Cf).V}{W} \tag{2}$$

$$Q = \frac{(Co - Cf).100}{Co} \tag{3}$$

Where V(L) is the solution volume, W(g) is the adsorbent amount, C_0 (mg/L) is the initial concentration of MO, and Cf (mg/L) is the MO concentration at equilibrium.

4.1 Adsorption isotherm studies

Adsorption isotherms provide information regarding the adsorbent capacity and the nature of the sorbent-solute interaction. Additionally, the isotherm constant values are essential to predict the maximum adsorption capacity, describing the affinity and the adsorbent surface properties. To describe the adsorption process of MO on activated charcoal that was obtained from the coconut shell, three traditional adsorption isotherms are used at 298 K: the Langmuir model [42], Freundlich model [43], and Brunauer–Emmett–Teller (BET) model [44].

4.2 Langmuir isotherm

The Langmuir Adsorption Isotherm is obtained once the adsorption is restricted by the formation of the monolayer when the adsorbate covers the adsorbent, it covers the monolayer and the process stops [45]. It is based on the hypothesis that states "all active adsorption centers are equivalent and the ability of a molecule to bind to the surface is independent whether or not there are nearby positions occupied" [46].

The linear representation of the Langmuir isotherm is represented by the Eq. (4):

$$\frac{C_e}{q_e} = \frac{1}{q_m K_L} + \frac{C_e}{q_m} \tag{4}$$

Where: q_m represents the maximum absorbent amount of monolayer (mg/g), K_L is the constant of Langmuir adsorption (L/mg) and is related to the free energy adsorption and temperature function. C_e is the concentration of the solution at equilibrium; q_e is the amount of adsorbate adsorbed at equilibrium per unit mass of the adsorbent (mg/g).

The essential characteristics of Langmuir isotherm can be expressed in terms of the equilibrium or separation parameter, called R_L factor, which is a dimensionless constant, Eq. (5) [47, 48]:

$$R_L = \frac{1}{1 + K_L C_o} \tag{5}$$

The R_L value indicates that the nature of adsorption is unfavorable ($R_L > 1$), linear ($R_L = 1$), favorable ($0 < R_L < 1$), or irreversible ($R_L = 0$). On the other hand, depending on the type of associated isotherm to Langmuir, the porosity characteristics associated with the material can be inferred [49].

4.3 Freundlich isotherm

The Freundlich adsorption isotherm model is an empirical equation that expressed the heterogeneity of the adsorbent material and multilayer coverage of adsorbate, its linear equation is represented by the Eq. (6)

$$ln\, q_e = ln\, K_F + \frac{1}{n}\, ln\, C_e \tag{6}$$

Where K_F is the Freundlich constant (mg/g) and is related to the adsorption capacity (bond strength), n is a parameter that measures the intensity of adsorption bond between the adsorbate and adsorbent.

4.4 Brunauer, Emmett Teller isotherm

The Brunauer, Emmett Teller (BET) isotherm model is an extension of the Langmuir model to the multilayer adsorption. It is not limited to the formation of monolayer, but the adsorption process is carried out until adsorbent surface is fully covered by a multimolecular or multilayer, which can be applied to both flat and convex surfaces [45]. The BET isotherm considers more real conditions and works under fewer assumptions, unlike Langmuir, which is why it is more successful to handle BET isotherm when it is required to reduce the surface area.

Its linearized form that allows direct graphic representation from the experimental data of the adsorption isotherm is presented in Eq. 7.

$$\frac{P}{V(Po - P)} = \frac{1}{V_m K} + \frac{K - 1}{V_m K} + \frac{P}{Po} \tag{7}$$

Where P is the pressure after the adsorption process; Po, the liquefaction of gas pressure; V is the volume adsorbed per gram of adsorbent; Vm, is the volume associated with the monolayer formation; K is a constant that depends on the temperature but does not depend on the surface coating [49].

To calculate the adsorbent surface area, the following equation was used:

$$A = \frac{V_m}{22400} N.\sigma.10^{-20} \tag{8}$$

Where: A is the adsorbent surface area (m^2/g), Vm is expressed in cm^3/g adsorbent, 22,400 is the molar volume in STP, N = Avogadro's Number, σ = adsorbate molecular area, in this case N_2 normally in $Amgstrom^2$ [45].

4.5 Thermodynamics of adsorption

The results of thermodynamic studies are useful to understand the viability of the adsorption process to obtain useful information regarding fundamental thermodynamic parameters of adsorption that are, such as the standard free energy change (ΔG^o). If the adsorption isotherms that were experimentally obtained are adequately described by the Langmuir's equation, then the thermodynamic parameters can be calculated by the Eqs. (9) and (10):

$$K = K_L.\gamma \tag{9}$$

$$\Delta G^o = -RTlnK \tag{10}$$

Where K is the dimensionless constant, K_L is the adsorption constant expressed in L/mol, γ is the amount of solvent in 1 kg of its weight (for water γ = 55.5 mol/L, water density 1 kg/L) [50, 51], R is the universal gas constant (8.31434 J/mol.K), and T is the absolute temperature.

5. Results and discussion

5.1 Optimal carbonization temperature

The charcoal obtained from the first carbonization at 450°C and 500°C, without being chemically activated, was subjected to adsorption tests with methyl orange solutions (200 mg/L and 400 mg/L), in order to determine which of these samples adsorbed more colorant. **Figure 3** shows that there is a higher percentage of MO adsorbed at 450°C rather than at 500°C. This difference in adsorption percentages between the charcoal obtained at 450°C and 500°C may be because the latter has a higher percentage of ash, which can interfere with the adsorption of the colorant, considering that these ashes are impurities that harm the adsorption process.

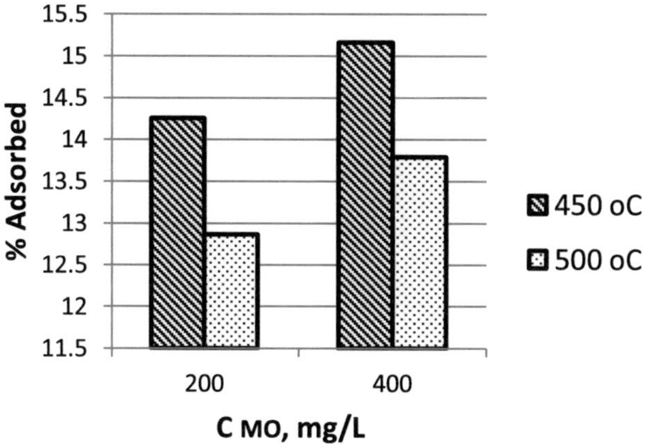

Figure 3.
Comparison of carbonization temperatures (heat treatment,) through methyl orange adsorption, C (MO concentration) at 450°C and 500°C.

5.2 Screening

Particles in the range of 150–250 μm [52] were selected for activation while using a standard analytical screen of frame and mesh. It was decided to sieve the particles obtained, in order to get a more homogeneous charcoal and that the difference in size between particles would not interfere with the adsorption process [34].

5.3 Comparison of charcoal obtained before and after activation

Activated charcoal with heat and chemical treatment showed a higher adsorption capacity with MO when compared with charcoal before the chemical activation process (only with heat treatment.) For the MO solutions of 200 and 400 mg/L, the adsorption increase was 140.6% and 51.3%, respectively, which indicates the effect of chemical activation with phosphoric acid in the adsorption process, see **Table 3**. This can also be compared with the work carried out by Carriazo et al. [53], where similar results are obtained.

It is observed an adsorption percentage increase which is possible because the chemical activation improved the surface area due to increased porosity [54]. Acid (phosphoric acid) impregnation oxidizes the charcoal's porous surface by increasing the hydrophilic locations on the surface [55]. Liquid-phase oxidations especially increase the concentration of carboxylic acids on the charcoal surface [56]. However, this is not evaluated in the present work, since the objective is to establish the adsorptive capacity of a material that is generated from a recycling product, which is why the presence of functional groups is not quantified, but it could be inferred about it, as it is indicated in the investigation carried out by Van Der Kamp et al. [57] and Bernal et al. [58]. By chemically activating the charcoal, the attraction between the adsorbent and the adsorbate increases, new bonds are formed between them, and it can be due to intermolecular forces. This type of initial physical adsorption occurs by weak forces and, generally, nonspecific forces, such as Van der Waals and the London dispersion forces [59–61]. On the other hand, the electrostatic interactions are common in the adsorption of activated charcoal due to the charges that are present when the acids or weak bases are ionized in an aqueous environment. Likewise, the presence of

Co MO (mg/L)	Charcoal (Treatment)	% Adsorbed	Adsorption variation, %
200	Non-activated	14.3	140.6
200	Activated	34.4	
400	Non-activated	15.2	51.3
400	Activated	23	

Table 3.
Comparison of the methyl orange adsorption capacity of the charcoal obtained at 450°C without activating (only with heat treatment) and activated charcoal with phosphoric acid; Co (initial concentration.) variation in the adsorption, percentage.

functional groups on the adsorbent surface generates the formation of dipoles with the adsorbate molecules due to differences in the electronic distributions. Aburub and Wuster [57] mention that specific interactions (ionic interactions, Keeson and Debye forces) are related to adsorption processes, which are directed through enthalpy, while nonspecific interactions (London forces) are related to processes that are carried through entropy [58].

5.4 Equilibrium experiments, adsorption isotherms

Adsorption isotherms are the tools used to predict the distribution of adsorbate molecules in the solid–liquid interface when the chemical equilibrium is reached. The MO adsorption isotherms on activated charcoal and the respective adjustments for the Langmuir model were obtained through the data shown in **Table 4** and represented in **Figure 4** where the straight line is reproduced with a coefficient of determination of 0.9822.

The shape of the isotherm adjusts to the Langmuir isotherm type one where a rapid increase in adsorption is observed while the pressure increases and stops when it reaches a limit value. This occurs because in this type of isotherm, the adsorption is restricted by the formation of the monolayer. This type of isotherm is present mostly in chemisorption processes [45]. In **Figure 5**, the monolayer formation is represented where the amount adsorbed increases with the MO concentration until it reaches a limit value, corresponding to the coating of a surface by a monolayer. First, there must be an adsorption–desorption equilibrium process, typical of physical adsorption. Once the monolayer formation is reached, comes the chemisorption process, which is predicted by this isotherm.

Ce/qe, g/L	qe, mg/g	Co (mg/L)	Ce (mg/L)	Q (%)
0.17	9.59	40	1.65	95.88
1.37	14.91	80	20.38	74.53
7.64	17.19	200	131.25	34.38
13.74	22.54	400	309.83	22.54

Table 4.
MO adsorption on activated charcoal; Co (initial concentration, mg/L), qe (adsorption capacity, mg/g), Q (removal efficiency), V (volume of MO solution, L).

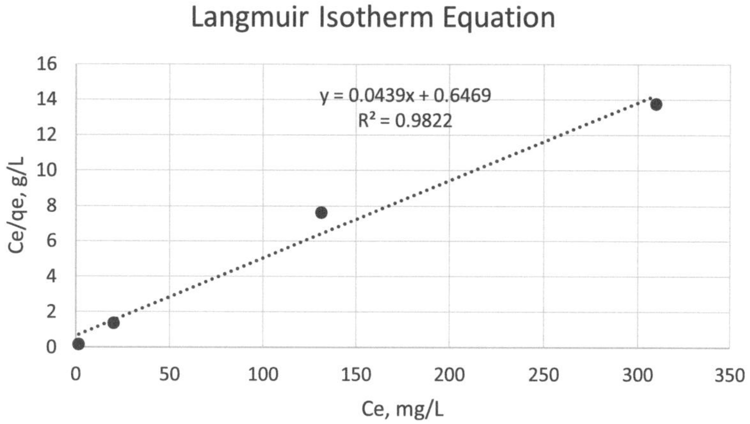

Figure 4.
Equation that represents the Langmuir isotherm, linearized, for the activated charcoal that is obtained from the coconut shell, moles adsorbed per gram of charcoal with respect to the final concentration; Ce: Final concentration of the solution (mg/L); qe: Amount of adsorbate adsorbed in the equilibrium per unit mass of the adsorbent (mg/g).

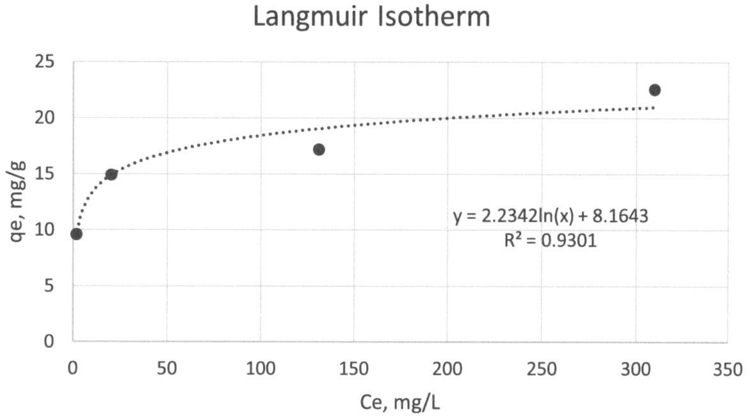

Figure 5.
Langmuir isotherm (type 1.) monolayer formation for activated charcoal from the coconut's endocarp; Ce: Final concentration of the solution; (mg/L) qe: mg adsorbed per gram of adsorbent.

The monolayer formation is reached when the adsorbed mass gets closer to 22.78 mg. A horizontal asymptote is observed when the moles adsorbed per gram of charcoal between the number of adsorption positions that are available in the surface approximate to 1. This type of isotherm is associated with microporous materials [62].

The constant values of the adsorption isotherms are shown in **Table 5**. The applicability of the isotherm equations is compared based on the coefficients of correlation R^2. The coefficient R^2 was lower in Freundlich rather than in Langmuir, which indicates that the MO adsorption in the charcoal activated adsorbent of the coconut shell results from adsorption in the monolayer.

The equilibrium parameter or separation called R_L factor obtained for the different initial concentrations of MO is shown in **Table 6**. The R_L value ($0 < R_L < 1$) indicates that the adsorption nature is favorable.

Langmuir model			Freundlich model		
q_m (mg/g)	K_L (L/mg)	R^2	K_F (mg/g)	n	R^2
22.78	0.0679	0.98225	8.99	0.1514	0.9653

Table 5.
Parameters of Langmuir and Freundlich isotherms, at 25°C.

Co (MO) (mg/L)	40	80	200	400
R_L	0.2692	0.1555	0.0686	0.0355

Table 6.
R_L factor (dimensionless) obtained for different initial concentrations of MO (mg/L).

5.5 BET isotherm, pore characteristics, and surface area

The graph obtained for the adsorption–desorption process for the nitrogen on activated charcoal from the coconut endocarp, applying the BET isotherm model, is shown below (**Figure 6**).

On the other hand, the values obtained for the BET isotherm with the activated charcoal were plotted, in order to evaluate its surface area, obtaining the graph shown below (**Figure 7**).

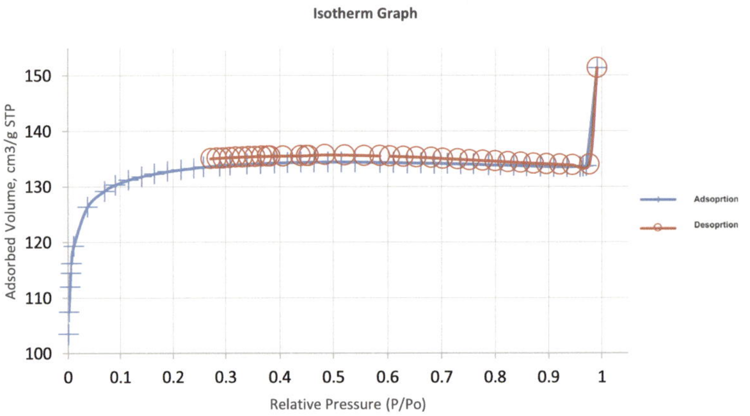

Figure 6.
BET isotherm representation, adsorption–desorption process for activated charcoal from the coconut endocarp; adsorbed volume (cm^3/g) STP based on the relative pressure (P/Po.)

The surface area of the activated charcoal was measured by using the Brunauer–Emmett–Teller (BET) model. The surface area according to BET is determined by applying the Eq. (8), and the value obtained of the specific surface area is 526 m^2/g. This value coincides with the values reported by [63] similar adsorbents.

Table 1 shows the results obtained for the surface area and pore volume using the BET model.

The total pore volume is estimated from the amount of nitrogen adsorbed at the highest relative pressure and the micropore volume is calculated from the nitrogen adsorption isotherm using the Dubinin-Radushkevich Equation [64].

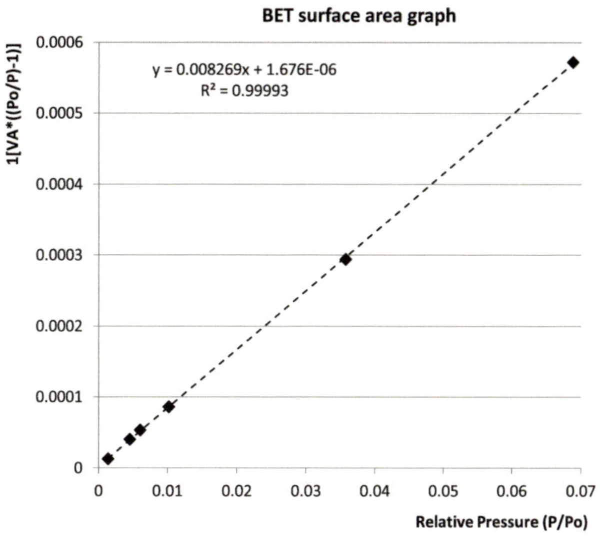

Figure 7.
BET isotherm representation, linearized, for activated charcoal from the coconut endocarp; 1/[VA((Po/P)-1)] based on the relative pressure (P/Po).*

The method used to determine the distribution of the pore size was the one from BJH [65], which was applied to the nitrogen adsorption data that was measured at 77 K in mesoporous materials. The results obtained match the ones reported by BJH (Barret-Joyner–Halenda), which estimate that the pore volume corresponds to a cylindrical pore volume (**Figure 8**) [65].

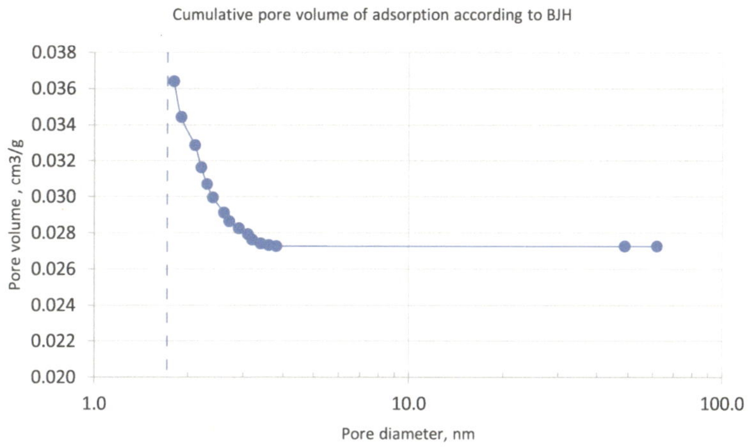

Figure 8.
Representation of pore volume (cm3/g) based on the pore diameter (nm), BJH model.

5.6 Thermodynamics of adsorption

The data that were obtained experimentally from the batch adsorption studies were analyzed using thermodynamic equations expressed before. **Table 7** shows the thermodynamic adsorption values.

K_L (L/mol)	ΔG° (kJ/mol)
22,215	−34.8

Table 7.
Thermodynamic values of MO adsorption on activated charcoal obtained from the coconut endocarp at 25°C.

The negative values of ΔG° at the studied temperature indicate the spontaneity of adsorption process, which reflects the affinity of the charcoal obtained toward the anionic dyes [66, 67]. The determined value is ΔG^o_{ads} = −34.8 kJ/mol, which is in the range from −20 kJ/mol to −40 kJ/mol. This probably indicates that physisorption and chemisorption are involved in this process [68]. It was taken into account another author for the calculation of K_L [69], where the ratio of the areas that the solvent and colorant occupy on the surface of the adsorbent is used, the value of ΔG^o_{ads} = −30.3 kJ/mol.

5.7 Effect of MO initial concentration on adsorption

The effect of MO initial concentration on adsorption by activated charcoal is shown in **Figure 9**. It can be seen that the equilibrium adsorption capacity increased almost linearly when the concentration of MO solution increased from 80 to 400 mg/l. This phenomenon can be explained in terms of interactions between the MO anionic molecule and the adsorbent. In this case, the amount of adsorbent is kept constant. When the MO amount per unit volume of solution increases, the ratio of the amount of MO ions in relation to the available adsorption site, at first, also increases and more MO ions in solution can be adsorbed by the activated charcoal. The observed behavior coincides with what was reported by [70].

The dissociation equilibrium of methyl orange is shown in **Figure 1**.

5.8 Morphology. Microstructure evolution of activated charcoal from the coconut shell

From the **Figures 10–12**, the morphology transformation of the coconut endocarp can be visualized. **Figure 10** shows the coconut endocarp without heat treatment.

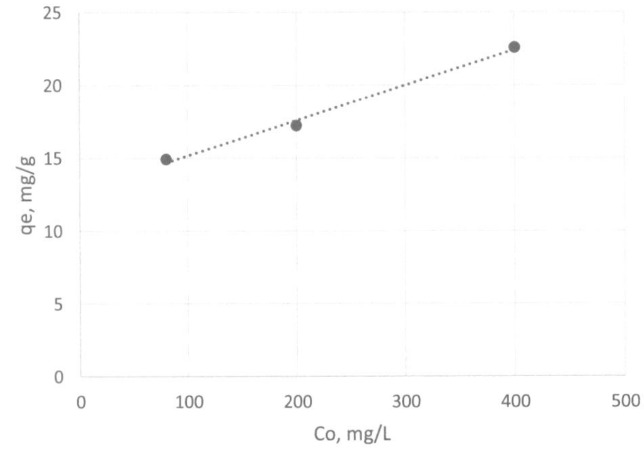

Figure 9.
Effect of MO initial concentration on activated charcoal adsorption.

Figure 11 shows the effect of carbonization on the coconut endocarp, and in **Figure 12**, the activated charcoal can be seen, which is the result from the chemical activation.

In **Figure 10**, the coconut endocarp without heat treatment, at different magnifications, is shown. An irregular morphology composed by fibers, fissures, and cavities is detected, which also indicates the potential of this material to generate activated charcoal, as referred by different authors [71–73]. Katesa et al. [74] concluded in their work that the porous properties of activated charcoal, including the surface area and pore volume, decreased while the carbonization temperature was increased. The lowest carbonization temperature at 250°C produced the activated charcoal with the highest porous properties for the activation temperature at 850° C and 60 and 120 minutes of activation time.

The higher magnification photomicrograph (**Figure 10d**) shows cavities of circular nature.

After carbonization (heat treatment,) an irregular external surface with cracks and non-spherical cavities around 0.1 μm is observed (**Figure 11**). As reported in the literature [71–74], the morphology obtained after heat treatment seems to be a result of the loss of moisture and volatile material that leaves the precursor structure; hence, leaving irregular cavities and transforming the structure of the coconut endocarp.

The activated charcoal shows great porosity where particles from 150 to 250 μm are heterogeneous, as well as their pores have different shapes and sizes, which favor the adsorption process, and it is in these pores where the adsorbate particles are housed, **Figure 12**.

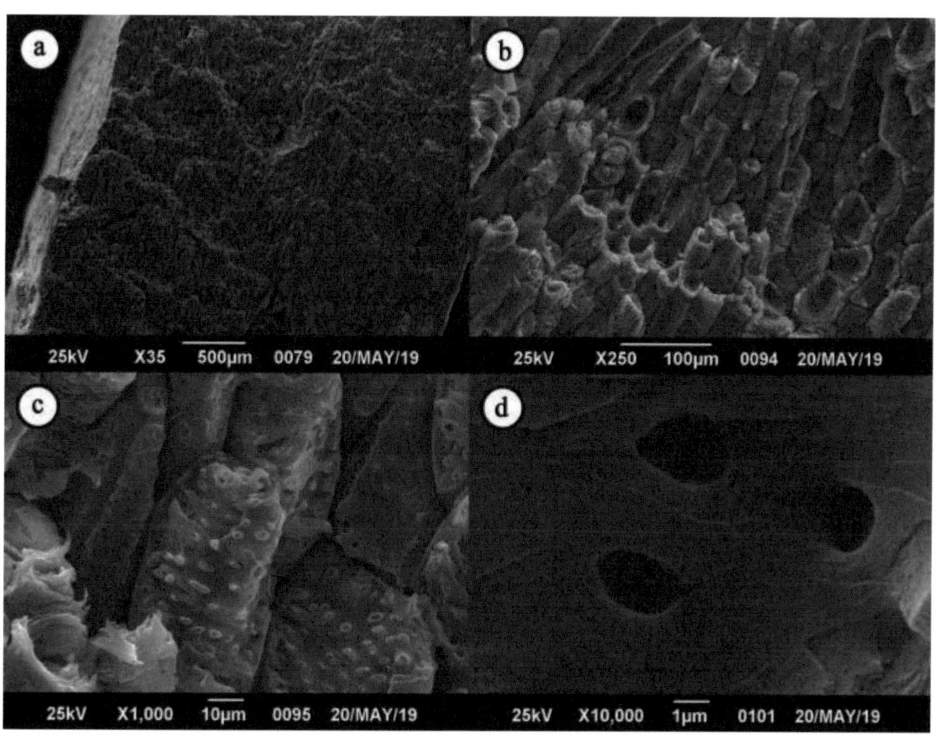

Figure 10.
Coconut endocarp with magnification of: a) 35, b) 250, c) 1000 y d) 10,000 times its size.

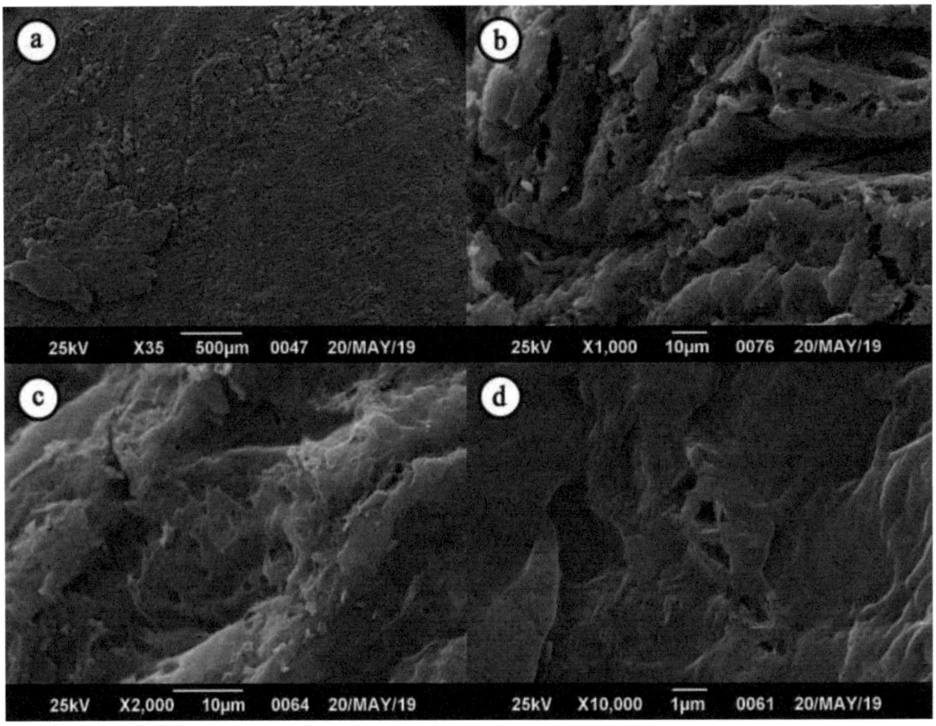

Figure 11.
Carbonized coconut endocarp at different magnifications: a) 35, b) 1000, c) 2000, and d) 10,000 times its size.

Representative images of the activated charcoal sample were selected in which can be observed the diversity in the size of the pore structure. **Figure 12c** shows pores of different sizes, **Figure 12d** shows smaller pore approaching 1000 nm.

Once the material was chemically activated (**Figure 12**), irregular particles, from 150 to 250 μm, and the development of great porosity were observed. This result confirms what is described in the literature [71–73], where it is mentioned that activating agents react with the carbonaceous chain to generate new pores or enlarge existing pores. Another possibility of the development of pores in activated charcoal is the occurrence of thermal stress in the structure of the precursor material that would lead to the formation of different cavities and fissures in the final product. Specifically, in the present study, phosphoric acid was used as an activating agent and expansion of the existing pores in the coconut endocarp was detected, as well as the formation of new macropores, mesopores, and micropores. In **Figure 12c**, macropores of different sizes are seen and in **Figure 12d**, macropores of around 1000 nm can be detailed.

The literature [71, 72] refers to the enormous effort in identifying precise methods and procedures to characterize activated charcoals and, specifically, its pore structure. Indirect techniques make it possible to correlate adsorption measurements with mathematical models and use this information to infer about the porous structure of activated charcoal. In this regard, the BET results of the present study show the presence of micropores with an average pore diameter around 1.7784 nm. On the other hand, the current computerized image analysis allows the conversion from 2D SEM (Scanning Electron Microscopy) to 3D. They open a window to improve the studies of porosity development and characterization of activated charcoal [75, 76], which is out of scope of the present study.

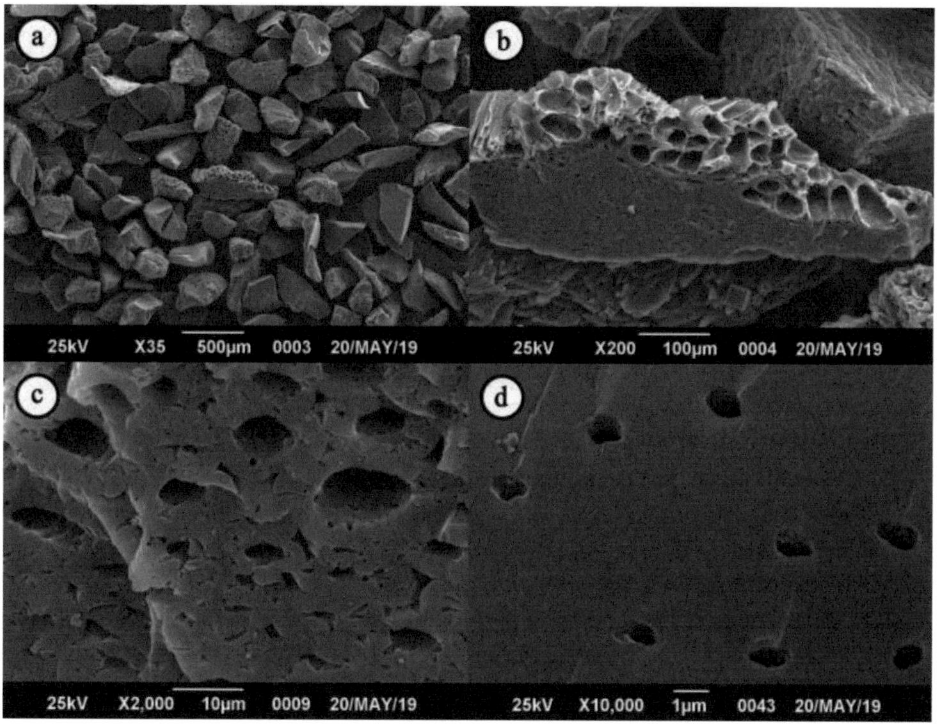

Figure 12.
Activated charcoal from the coconut endocarp with a magnification of: a) 35, b) 200, c) 2000, and d) 10,000 times its size.

It was also possible to determine through the EDS analysis (**Figure 13** and **Table 8**) that the charcoal obtained has the presence of charcoal, oxygen, and potassium. Although, part of these elements come from the coconut material, the oxygen present is the result of the carbonization and activation of the sample in the presence of air, and potassium is present in the endocarp, pulp, and water of the coconut [77].

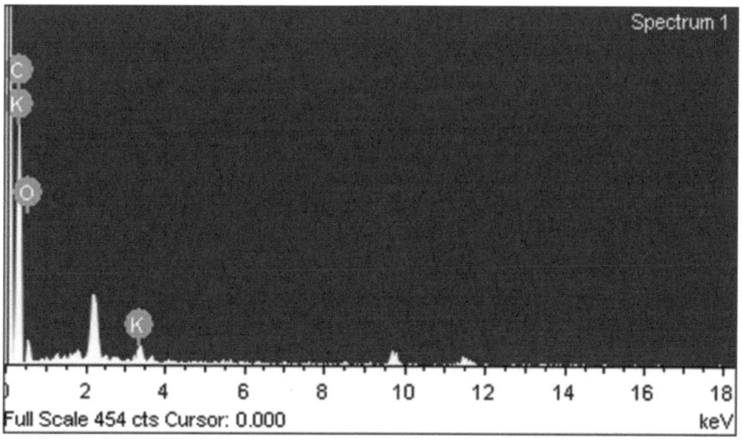

Figure 13.
Energy dispersive X-ray spectroscopy (EDS) of activated charcoal from coconut shell. Source: Energy dispersive X-ray spectroscopy, adapted to SEM.

Element	Mass %	Atomic %
C	75.58	81.03
O	22.98	18.50
K	1.44	0.47
Total	**100.00**	**100.00**

Table 8.
Composition of activated charcoal from coconut shell through energy dispersive X-ray spectroscopy (EDS).

6. Conclusions

This study revealed the potential of the coconut endocarp as a good precursor for the preparation of activated charcoal. The porous structure developed with a surface area, a total pore volume, and an average pore diameter, according to BET, of 526 m^2/g, 0.2341 cm^3/g, and 1.7784 nm, respectively, improves the adsorption process and indicates the presence of micropores. The equilibrium studies showed that the isotherm Langmuir model matches the adsorption data, meaning that the MO colorant adsorption forms a monolayer on the CACC. The adsorption capacity at equilibrium increased almost linearly when the concentration of MO solution increased from 80 to 400 mg/L. The equilibrium parameter R_L, obtained for the different initial concentrations of MO, indicates that the nature of adsorption is favorable. The thermodynamic parameters, that is, the negative values of Gibbs energy at the studied temperature, indicate, on one hand, the spontaneity of the adsorption process, reflecting the affinity of the charcoal obtained toward the anionic dyes, and on the other hand, that in the process are probably involved the physisorption and chemisorption. The SEM morphology analysis of CACC showed the presence of macropores and mesopores, which are characteristic of the activated charcoal. The results indicate that the activated charcoal obtained could be used as an alternative low-cost adsorbent in the MO removal from effluents in aqueous solutions.

Acknowledgements

The authors thank Universidad Metropolitana, Universidad Simón Bolívar, and PDVSA-INTEVEP for allowing this research to be carried out in their laboratories.

Declaration of conflict of interests

The authors who appear in the article declare that there is no potential conflict of interests related to it.

Author details

Isabel Cristina Páez-Pumar Romer†, Isabella Victoria Plazola Santana,
Rosa María Rodríguez Bengoechea and Miguel Manuel Pérez Hernández*
Universidad Metropolitana, Venezuela, Caracas

*Address all correspondence to: mperez@unimet.edu.ve

† These authors contributed equally.

References

[1] Tsuboy MS, Angeli JPF, Mantovani MS, Knasmueller S, Umbuzeiro GA, Ribeiro LR. Genotoxic, mutagenic and cytotoxic effects of the commercial dye CI disperse blue 291 in the human hepatic cell line HepG2. Toxicology In Vitro. 2007; **21**(8):1650-1655. DOI: 10.1016/j.tiv.2007.06.020

[2] Vinitnantharat S, Chartthe W, Pinisakul A. Toxicity of reactive red 141 and basic red 14 to algae and Waterfleas. Water Science and Technology. 2008; **58**(6):1193-1198. DOI: 10.2166/wst.2008.476

[3] Anjaneyulu Y, Sreedhara-Chary N, Suman-Raj S. Decolourization of industrial effluents available methods and emerging technologies a review. Rev. Environ. Sci. Technol. 2005;**4**:245-273. DOI: 10.1007/s11157-005-1246-z

[4] Khelifi O, Nacef M, Affoune AM. Nickel (II) adsorption from aqueous solutions by physic-chemically modified sewage sludge. Iranian journal of chemistry and chemical engineering. 2018;**37**(1):73-87. DOI: 10.30492/IJCCE.2018.29994

[5] Belaid KD, Kacha S. Etude cinétique et thermodynamique de l'adsorption d'un colorant basique sur la sciure de bois. Journal of Water Science. 2011, 2011;**24**(2):131-144. DOI: 10.7202/1006107ar

[6] Larakeb M, Youcef L, Achour S. Etude comparative de l'élimination du Zinc par adsorption sur la goethite et sur la bentonite de Maghnia. LARHYSS Journal. 2014;**19**(3):87-100. DOI: 10.7202/1006107ar

[7] Beijer E, Björlenius B, Shaik S, Lindberg RH, Brunström B, Brandt I. Removal of pharmaceuticals and unspecified contaminants in sewage treatment effluents by activated carbon filtration and ozonation: Evaluation using biomarker responses and chemical analysis. Chemosphere. 2017;**176**: 342-351. DOI: 10.1016/j.chemosphere.2017.02.127

[8] Cazetta AL, Vargas AMM, Nogami EM, Kunita MH, Guilherme MR, Martins AC, et al. NaOH-activated carbon of high surface area produced from coconut shell: Kinetics and equilibrium studies from the methylene blue adsorption. Chemical Engineering Journal. 2011;**174**:117-125. DOI: 10.1016/j.cej.2011.08.058

[9] Arami M, Limaee NY, Mahmoodi NM, Tabrisi NS. Removal of dyes from colored textile wastewater by orange peel adsorbent: Equilibrium and kinetic studies. Journal of Colloid and Interface Science. 2005;**288**(2): 371-376. DOI: 10.1016/j.jcis.2005.03.020

[10] Khelifi O, Nacef M, Affoune AM. Biosorption of nickel (II) ions from aqueous solutions by using chicken eggshells as low-cost biosorbent. Algerian Journal of Environmental Science and Technology. 2016;**2**(1): 12-16. ISSN: 2437-1114

[11] Hazourli S, Ziati M, Hazourli A, Cherifi M. Valorisation d'un résidu naturel ligno-cellulosique en charbon actif - exemple des noyaux de dattes. In: Revue des énergies renouvelables. Algeria: ICRESD-07 Tlemcen; 2007. pp. 187-192

[12] Pekkuz H, Uzun İ, Güzel F. Kinetics and thermodynamics of the adsorption of some dyestuffs from aqueous solution by poplar sawdust. Bioresource

Technology. 2008;**99**(6):2009-2017. DOI: 10.1016/j.biortech.2007.03.014

[13] Li W, Yang K, Peng J, Zhang L, Guo S, Xia H. Effects of carbonization temperatures on characteristics of porosity in coconut shell chars and activated carbons derived from carbonized coconut shell chars. Industrial Crops and Products. 2008;**28**:190-198. DOI: 10.1016/j.indcrop.2008.02.012

[14] Mittal A, Malviya A, Kaur D, Mittal J, Kurup L. Studies on the adsorption kinetics and isotherms for the removal and recovery of methyl Orange from wastewaters using waste materials. Journal of Hazardous Materials. 2007; **148**(1–2):229-240. DOI: 10.1016/j.jhazmat.2007.02.028

[15] Sakr F, Sennaoui A, Elouardi M, Tamimi M, Assabbane A. Étude de l'adsorption du Bleu de Méthylène sur un biomatériau à base de Cactus. J. Mater. Environ. Sci. 2015;**6**(2):397-406. ISSN: 2028-2508 CODEN: JMESCN

[16] Ocampo-Perez R, Leyva-Ramos R, Mendoza-Barron J, Guerrero-Coronado RM. Adsorption rate of phenol from aqueous solution onto organobentonite: Surface diffusion and kinetic models. Journal of Colloid and Interface Science. 2011;**364**(1):195-204. DOI: 10.1016/j.jcis.2011.08.032

[17] Bernal V, Giraldo L, Moreno-Piraján JC. Adsorción de acetaminofén sobre carbones activados a diferente pH. Entalpía y entropía del proceso. Revista Colombiana de Quimica. 2018;**47**(2): 54-62. DOI: 10.15446/rev.colomb.quim.v47n2.68213

[18] Valentina B, Liliana G, Moreno Juan C. Caracterización textural y química de carbones activados preparados a partir de cuesco de palma africana (*Elaeis guineensis*) por activación química con CaCl2 y MgCl2. Revista Colombiana de Quimica. 2018;**47**(2): 54-62. DOI: 10.15446/rev.colomb.quim.v47n2.68213

[19] Molina-Sabio M, Rodríguez-Reinoso F. Role of chemical activation in the development of carbon porosity. Colloid Surface. 2004;**241**:15-25. DOI: 10.1016/j.colsurfa.2004.04.007

[20] Boehm HP. Surface oxides on carbon and their analysis: A critical assessment. Carbon. 2002;**40**:145-149. DOI: 10.1016/S0008-6223(01)00165-8

[21] Guo J, Chong Lua A. Textural and chemical properties of adsorbent prepared from palm shell by phosphoric acid activation. Materials Chemistry and Physics. 2003;**80**:114-119. DOI: 10.1016/S0254-0584(02)00383-8

[22] Enio S, Eustaquio A, Rafael R, Fidel R. Características agronómicas y productivas de progenies de cocotero alto por enano, falcón-venezuela. Agronomía Tropical. 2003; **53**:17-32

[23] Giraldo L, Moreno-Piraján JC. Monolitos de Carbón Activado a partir de Cáscara de coco e impregnación con níquel y cobre. Revista Colombiana de Quimica. 2008;**37**(3):355-370. Recuperado de: http://www.scielo.org.co/pdf/rcq/v37n3/v37n3a09.pdf

[24] Mojica LC, Ramírez WM, Rincón NG, Blanco DA, et al. Síntesis de carbón activado proveniente de semillas de Eucalipto por activación física y química. Bogotá, Colombia: Universidad de Los Andes; 2012

[25] De la Hoz M, Sarmiento J. Carbón Activado: Alternativa para la Purificación de Gases de Combustión [Trabajo de Grado]. Venezuela: Universidad Metropolitana; 2018

[26] Academic. Naranja de metilo. 2010. Recuperado de: https://esacademic.com/dic.nsf/eswiki/565589

[27] Ramírez Llamas LA, Jacobo Azuara A, Martínez Rosales JM. Adsorción del naranja de metilo en solución acuosa sobre hidróxidos dobles laminares. Acta Universitaria. 2015;**25**(4):25-34. DOI: 10.15174/au.2015.778

[28] Ocholi OJ, Gimba CE, Ndukwe GI, Turoti M, Abechi SE, Edogbanya PRO. Effect of time on the adsorption of methylene blue, methyl Orange and indigo carmine onto activated carbon. IOSR Journal of Applied Chemistry (IOSR-JAC). 2016;**9**(9) Ver. I:55-62. DOI: 10.9790/5736-0909015562

[29] Zhe-Ming N, Sheng-Jie X, Li-Grng W, Fang-Fang X, Guo-Xiang P. Treatment of methyl orange by calcined layered double hydroxides in aqueous solution: Adsorption property and kinetic studies. Journal of Colloid and Interface Sciences. 2007;**316**(2):284-291. DOI: 10.1016/j.jcis.2007.07.045

[30] Kron DA, Holland BT, Wipson R, Maleke C, Stein A. Anion exchange properties of a mesoporous Aluminophosphate. Langmuir. 1999; **15**(23):8300-8308. DOI: 10.1021/la990553r

[31] Cisneros A, Fernandes E. Preparación de carbón activado a partir de residuos de torrefaccionado y carbonizado alcalino de biomasa vegetal (Equisetum Giganteum) [Tesis de Grado]. Universidad Metropolitana, Caracas, Venezuela; 2017

[32] Lopez-Ramon MV, Stoeckli F, Moreno-Castilla C, Carrasco-Marin F. On the characterization of acidic and basic surface sites on carbons by various techniques. Carbon. 1999;**37**:1215-1221

[33] Coasne B, Galarneau A, Gerardin C, Fajula F, Villemot F. Validity of the t-plot method to assess microporosity in hierarchical micro/mesoporous materials. Langmuir. 2014;**30**:13266-13274

[34] Páez-Pumar I, Plazola I. Obtención y evaluación de carbón activado a partir del endocarpio de coco (cocos nucifera) para la remoción de compuestos minerales en agua potable [Tesis de Grado]. Caracas, Venezuela: Universidad Metropolitana; 2019

[35] Subasioglu T, Bilkay IS. Determination of biosorption conditions of methyl Orange by Humicola fuscoatra. Journal of Scientific and Industrial Research. 2009;**68**(12): 1075-1077

[36] Tchuifon DR, Anagho SG, Njanja E, Ghogomu JN, Ndifor-angwafor NG, Kamgaing T. Equilibrium and kinetic modelling of methyl orange adsorption from aqueous solution using rice husk and egussi peeling. International Journal of Chemical Sciences. 2014;**12**(3): 741-761. ISSN: 0972-768X

[37] Saha TK, Bhoumik NC, Mahmooda SK, Hideki Ichikawa GA, Fukumori Y. Adsorption of methyl Orange onto chitosan from aqueous solution. J. Water Resource and Protection. 2010;**2**:898-906. DOI: 10.4236/jwarp.2010.210107

[38] Aboua KN, Soro DB, Diarra M, DIBI K, N'Guettia KR, Traore KS. Étude de l'adsorption du colorant orange de méthyle sur charbons actifs en milieu aqueux: influence de la concentration de l'agent chimique d'activation. Afrique Science. 2018; **14**(6):322-331

[39] Arenas IA, López JL. Espectrofotometría de absorción. Curso: Métodos de Laboratorio, Instituto de

Biotecnología, Universidad Nacional Autónoma de México; 2004. Recuperado de: http://www.ibt.unam.mx/computo/pdfs/met/espectrometria_de_absorcion.pdf

[40] Haris MR, Sathasivam K. The removal of methyl red from aqueous solution using banana pseudostem fibers. American Journal of Applied Sciences. 2009;**6**(9):1690-1700

[41] Gamarra CG, La Rosa-Toro A. Decoloración del anarajado de metilo empleando el sistema fenton. Rev Soc Quím Perú. 2014;**80**(1):24-34. ISSN: 1810-634X

[42] Zhang PK, Wang L. Extended Langmuir equation for correlating multilayer adsorption equilibrium data. Separation and Purification Technology. 2010;**70**(3):367-371. DOI: 10.1016/j.seppur.2009.10.007

[43] Jeppu GP, Clement TP. A modified Langmuir-Freundlich isotherm model for simulating pH-dependent adsorption effects. Journal of Contaminant Hydrology. 2012;**129-130**:46-53. DOI: 10.1016/j.jconhyd.2011.12.001

[44] Naderi M. Chapter Fourteen - Surface Area: Brunauer–Emmett–Teller (BET). In: Tarleton S, et al. Progress in Filtration and Separation. UK: Academic Press; 2015, pp. 585-608, ISBN 9780123847461, DOI: 10.1016/B978-0-12-384746-1.00014-8

[45] Shaw D. Química de superficie y coloides. Madrid, España: Ediciones, Alhambra, S.A; 1977

[46] Lazo JC, Navarro A, Sun-Kou MR, Llanos BP. Síntesis y caracterización de arcillas organofílicas y su aplicación como adsorbentes del fenol. Rev. Soc. Quim. Perú. 2008;**74**(1):3-19. Disponible en: http://www.redalyc.org/articulo.oa?id=371937608002

[47] Ahmad M, Rajapaksha AU, Lim JE, Zhang M, Bolan N, Mohan D, et al. Biochar as a sorbent for contaminant management in soil and water. Chemosphere. 2014;**99**:19-33. DOI: 10.1016/j.chemosphere.2013.10.071

[48] Ramírez A, Anyi P, et al. Preparación de carbón activado a partir de residuos de palma de aceite y su aplicación para la remoción de colorantes. Revista Colombiana de Quimica. 2017;**46**(1):33-41. DOI: 10.15446/rev.colomb.quim.v46n1.62851

[49] Castellan G. Fisicoquimica. Segunda edición. Editorial Pearson. Míxico: Addison Wesley Iberoamericana, S.A; 1987

[50] Zhou X, Zhou X. The unit problem in the thermodynamic calculation of adsorption using the Langmuir equation. Chemical Engineering Communications. 2014;**201**(11):1459-1467. DOI: 10.1080/00986445.2013.818541

[51] Liu Y. Is the free energy change of adsorption correctly calculated? Journal of Chemical & Engineering Data. 2009;**54**(7):1981-1985. DOI: 10.1021/je800661q

[52] Sarmiento C, Sánchez J, García C, Rincón Y, Benítez A, Ramírez J. Preparación de carbón activado mediante la activación química de carbón mineral [Trabajo de grado]. Maracaibo, Venezuela: Universidad del Zulia; 2004

[53] Carriazo JG, Saavedra MJ, Molina MF. Propiedades adsortivas de un carbón activado y determinación de la ecuación de Langmuir empleando materiales de bajo costo. Educ. quím. 2010;**21**(3):224-229. © Universidad

Nacional Autónoma de México, ISSN 0187-893-X Publicado en línea el 3 de mayo de 2010, ISSNE 1870-8404

[54] Inamuddin AMA, Mohammad A. Organic Pollutants in Wastewater I: Methods of Analysis, Removal and Treatment. Vol. 29. India. USA. Capítulo 4: Publicado por Materials Research Forum LLC; 2018. pp. 101-103. ISBN 194529163X, 9781945291630

[55] Wenzhong S, Zhijie L, Yihong L. Surface chemical functional groups modification of porous carbon. Recent Patents on Chemical Engineering. 2008; **1**:27-40. DOI: 10.2174/ 2211334710801010027

[56] Figueiredo JL, Pereira MFR, Freitas MMA, Orfao JJM. Modification of the surface chemistry of activated carbons. Carbon. 1999;**37**(9):1379-1389. DOI: 10.1016/S0008-6223(98)00333-9

[57] Van Der Kamp KA, Qiang D, Aburub A, Wurster DE. Modified Langmuir-like model for modeling the adsorption from aqueous solutions by activated carbons. Langmuir. 2005; **21**(1):217-224. DOI: 10.1021/la040093o

[58] Bernal V, Liliana Giraldo L, Moreno-Piraján JC. Adsorción de acetaminofén sobre carbones activados a diferente pH. Entalpía y entropía del proceso. Revista Colombiana de Quimica. 2018;**47**(2): 54-62. DOI: 10.15446/rev.colomb.quim. v47n2.68213

[59] De Boer JH, Custers JFH. Adsorption by van der Waals forces and surface structure. Physica. 1937;**4**(10): 1017-1024. DOI: 10.1016/S0031-8914 (37)80199-8

[60] Kong L, Adidharma H. A new adsorption model based on generalized van der Waals partition function for the description of all types of adsorption isotherms. Chemical Engineering Journa l. 2019;**375**:122112. DOI: 10.1016/j.cej. 2019.122112

[61] Fulazzaky MA. Study of the dispersion and specific interactions affected by chemical functions of the granular activated carbons. Environme ntal Nanotechnology, Monitoring & Management. 2019;**12**:100230. DOI: 10.1016/j.enmm.2019.100230

[62] Knözinger E, Schüth F, Weitkamp J. Handbook of heterogeneous catalysis. 2., completely revised and enlarged edition. Vol. 1–8. Angewandte Chemie International Edition. 2009;**48**(19): 3390-3391. DOI: 10.1002/ anie.200901598

[63] Yusoff S, Kamari A, Putra W, Ishak C, Mohamed A, Hashim N, et al. Removal of Cu (II), Pb (II) and Zn (II) ions from aqueous solutions using selected agricultural wastes: Adsorption and characterisation studies. Journal of Environmental Protection. 2014;**5**: 289-300. DOI: 10.4236/jep.2014.54032

[64] Castro JB, Bonelli PR, Cerrella EG, Cukierman AL. Phosphoric acid activation of agricultural residues and bagasse from sugar cane: Influence of the experimental conditions on adsorption characteristics of activated carbons. Industrial and Engineering Chemistry Research. 2000;**39**(11):4166-4172. DOI: 10.1021/ie0002677

[65] Barrett EP, Joyner LG, Halenda PP. The determination of pore volume and area distributions in porous substances. I. Computations from nitrogen isotherms. Journal of the American Chemical Society. 1951;**73**:373-380. DOI: 10.1021/ja01145a126

[66] Özcan M, Solmaz R, Kardaş Dehri I. 2008 adsorption properties of

barbiturates as green corrosion inhibitors on mild steel in phosphoric acid colloids and surfaces. Physicochem. Eng. Aspects. 2008;**325**:57-63. DOI: 10.1016/j.colsurfa.2008.04.031

[67] Milonjic SK. A consideration of the correct calculation of thermodynamic parameters of adsorption. Journal of the Serbian Chemical Society. 2007;**72**(12): 1363-1367. DOI: 10.2298/JSC0712363M

[68] Saxena A, Prasad D, Haldhar R, Singh G, Kumar A. Use of Sida cordifolia extract as green corrosion inhibitor for mild steel in 0.5M H2SO4. Journal of Environmental Chemical Engineering. 2018;**6**(1):694-700. DOI: 10.1016/j.jece.2017.12.064

[69] Soldatkina L, Zavrichko M. Equilibrium, kinetic, and thermodynamic studies of anionic dyes adsorption on corn stalks modified by Cetylpyridinium bromide. Colloids and Interfaces. 2018, 2018;**3**(1):4. DOI: 10.3390/colloids3010004

[70] Huang ZN, Wang X, Yang DS. Adsorption of Cr (VI) in wastewater using magnetic multi-wall carbon nanotubes. Water Science and Engineering. 2015;**8**(3):226-232. DOI: 10.1016/j.wse.2015.01.009

[71] Budi E, Nasbey H, Yuniarti BDP, Nurmayatri Y, Fahdiana J, Budi AS. Pore structure of the activated coconut shell charcoal carbon. In: AIP Conference Proceedings 1617. Indonesia: AIP Publishing LLC; 2014. pp. 130-133. DOI: 10.1063/ 1.4897121

[72] Wei L, Yang K, Peng J, Libo Z, Sheughui G, Hongying X. Effects of carbonization temperatures on characteristics of porosity in coconut Shell chars and activated carbons derived from carbonized coconut Shell chars. Industrial Crops and Products. 2008;**28**(2):190-198. DOI: 10.1016/j.indcrop.2008.02.012

[73] Xinying W, Danxi L, Wei L, others. Optimization of mesoporous activated carbon from coconut shells by chemical activation with phosphoric acid. BioResources. 2013;**8**(4):6184-6195. DOI: 10.15376/biores.8.4.6184-6195

[74] Katesa J, Junpiromand S, Tangsathitkulchai C. Effects of carbonization temperature on porous properties of coconut Shell based activated carbon. Journal of Science and Technology. 2013;**20**(4):269-278. Corpus ID: 98597508

[75] Achaw O-W. A Study of the Porosity of Activated Carbons Using the Scanning Electron Microscope, Scanning Electron Microscopy, Viacheslav Kazmiruk. Rijeka: IntechOpen; 2012. DOI: 10.5772/36337. Available from: https://www.intechopen.com/chapters/30949

[76] Ab. Jabal SN, Seok YB, Wee FH. Carbon composition, surface porosities and dielectric properties of coconut Shell powder and coconut Shell activated carbon composites. ARPN Journal of Engineering and Applied Sciences. 2016; **11**(6):3832-3837. Corpus ID: 136667440. ISSN 1819-6608

[77] Coconut water is it, F.A.O. http://www.fao.org/News/2000/000902-e.htm. Última visita, 15/03/2020; 14:30 horas

Chapter 3

Sorption Isotherms and Some Functional Properties of Cowpea Varieties Flour

Issoufou Amadou

Abstract

In sub-Sahara, preservation of processed cowpea flour remained a challenge, and there are no standard isotherm conditions for drying cowpea flour. This study aims to define the optimum isotherm conditions for cowpea flour and assess their functional properties. Adsorption isotherms of three varieties of cowpea at temperatures 30, 40, and 50°C and in each case with six different applications depending on the constant relative humidity of the medium were executed. Water and oil absorption capacities including swelling index were determined. Results show that water content at equilibrium is inversely proportional to the temperature, and at the same temperature, the water content increases when water activity augments. The adsorption isotherms are of type II according to the fitted BET and GAB models. The absorption capacities ranged from 1.06 ± 0.01, 1.08 ± 0.02, and 1.09 ± 0.01(mL/g), respectively, for CS133, CS032, and control. However, the swelling index was significantly separated ($P < 0.05$). The adsorption isotherm curve of the sample CS032 at 50°C shows a stronger correlation ($R^2 = 0.9274$) than the other varieties regardless of the mathematical isotherm model used. It can be concluded that depending on some functional properties of cowpea variety flour, these varieties seemed to behave separately *vis-a-vis* their sorption isotherm.

Keywords: adsorption isotherm, cowpea varieties flour, GAB and BET model, swelling index

1. Introduction

The *Vigna unguiculata* (L) Walp, Cowpea, is a legume of genus *Vigna*, family *Fabaceae*, subfamily *Faboideae* [1, 2]. Although, cowpea plants are grown globally, it is suggested that they originated from Africa [3, 4]. Currently, West Africa is the major producer of cowpeas globally, of which 80% is from Central Africa [5]. Similar to beans, cowpeas are highly nutritious, hence a source of food and income globally. Cowpeas contain 17–42% protein, 1.4% fat, and 35–61% carbohydrate [6]. Compared to cereals, cowpeas are 2–3 times richer in amino acids, such as lysine, thiamine and riboflavin, and

 IntechOpen

carbohydrate; which make it also an energetic food [7]. Nowadays, cowpea farming has reached a level of important yielding in the area of its potential production. However, to make this important production profitable to the farmers, processing into cowpea flour as such is necessary, to improve the quality and profitability. Indeed, to stabilize cowpea flour, to make it competitive to other imported flour (cereals; flour), there was a need to study the drying conditions of cowpea flour [8]. Cowpeas have been a nutritious legume and the new trend in developing countries such as Sub-Saharan Africa representing about 95% of world cowpea production [9].

Drying food product such as cowpea flour is an essential step to evaluate its hygroscopic character, which indicates the affinity that the food product may have with its surrounding environment. Sorption isotherms represent the interrelation between the activity of water and the water content of food at a constant temperature. The most used technique to preserve the quality of food is to reduce the water activity to a sufficiently low level. Obtaining the sorption isotherm is essential in determining the moisture level at which microbial growth and mycotoxin production is inhibited during storage [10–12]; in addition, it predicts the speed and intensity of chemical and enzymatic reactions [11]. Drying and storage of industrial or artisanal processing operations required the know-how of the nature of water-substrate interactions or sorption isotherms [13]. When exposed to high temperatures, the functional properties of cowpea flours are affected [14–16]. This further affects the downstream processing of cowpea flour for food products. Examples of cowpea flour food products include spaghetti, couscous, porridge, infant donuts, dough, and infant flour as a dietary supplement [6, 17].

Biochemical components of cowpea varieties are expected to impact the properties of its sample, in which the hydrophilic behavior of protein has an effect on water-holding capacity. Then, the most important energy reserve, the carbohydrates, have also significant influence on the physicochemical properties of flour from cowpea varieties [1, 18]. Furthermore, the critical factors that influence the functional properties on downstream processing of food-legume products include oil and water absorption capacity [6]. The influence on the functional properties varies according to the biochemical properties of the legume, the phenomenon is likely to be even more distinct in cowpeas, especially from diverse varieties [11]. Therefore, after drying, it is relevant to assess the cowpea flour hydrophilicity (the affinity for water) and the sorption isotherm conditions (the temperature at with the flour absorbs water). The consequences of not assessing these parameters are that when the flour is not well dried, the downstream process is affected. Furthermore, the presence of moisture favors bacterial and fungal (such as aflatoxin) growth [10–12]. However, until now, the optimum isotherm conditions for drying cowpea flour are not known. The present work aimed to study the conditions of sorption isotherm for drying cowpea flour. The sorption isotherm conditions and their effect on functional properties were examined against flour from three varieties of cowpea plants. The findings will be vital in establishing a standard protocol for preserving cowpea flour, which will improve the processing of cowpea-based food products.

2. Materials and methods

2.1 Materials

Three cowpea dried seed samples were used in the study of which two seed samples (CS133 and CS032) were obtained from Dan Dicko Dankoulodo University of

Maradi (UDDM), Cowpea Square research project, Niger, whereas the control sample was obtained from the Maradi city, Niger.

2.2 Methods

2.2.1 Preparation of samples

First, the cowpea seed samples were sorted to remove foreign matter. Thereafter the seed samples were soaked in potassium hydroxide solution for 24 hours, dehulled, and dried at 90 ± 2°C for four (4) hours, followed by milling. The resulting flours were sifted using a 250 μm sieve. The sifted flours were collected in an air-free polyethylene bag, sealed, and stored for subsequent experiments.

2.2.2 Physico-functional properties

The moisture content of the samples was carried out by the methods of AOAC [19]. Both for oil and water (OAC/WAC) absorption capacities of seed samples, flours (1 g) were mixed into 10 mL, centrifuged, and then stirred with *Balanites aegyptiaca* oil or distilled water, respectively. Slurries were centrifuged at 3000 × *g* for 10 min (80–2 15/20 mL Electronic Lab Centrifuge Machine, Jiangsu, China), and the oil or water released after centrifugation was massed and expressed as (mL/g and g/g) OAC/WAC capacities, respectively, according to Sofi et al. [20] with some modifications.

The swelling index was obtained by taking a 1 g of cowpea flour sample in pre-weighed centrifuge tubes with distilled water (10 mL), vortexed; then incubated at room temperature on the shelf for 24 hours. The noted marked volume after 24 hours was considered as the total volume of flour sample and expressed as followed:

$$SV = \frac{TV}{W_i} \tag{1}$$

SV: swelling index (g/cm^3).
TV: Total volume (g).
Pi: Initial sample weight (g).

2.2.3 Determination of sorption isotherms

Equilibrium moisture content (EMC) by the static gravimetric method was used to evaluate moisture desorption isotherms at different water activity levels. In brief, cowpea flour samples (in triplicate) were placed in a desiccator, each containing a 250 mL solution of sulfuric acid at different relative humidity (10–90%) and preserved at the temperatures of 30, 40, and 50°C [21]. Periodically, samples were weighed precisely until no significant variation in weight was detected or till equilibrium was reached [22]. The EMC was calculated considering samples' initial moisture contents as described in the following:

$$EMC = \frac{w_f - w_i + \left[\frac{\%H_2O}{100} w_i\right]}{w_i\left[\frac{100 - \%H_2O}{100}\right]} \tag{2}$$

Models	Equations	Parameters
GAB [23]	$\frac{EMC}{M_m} = \frac{ABa_w}{[1-Ba_w)(1-Ba_w+ABa_w)]}$ Rearranged equation into the second-degree polynomial form $\frac{a_w}{EMC} = A_1a_w^2 + A_2a_w + A_3$	A_1 = [B/Mm] [(1/A) – 1] A_2 = (1/Mm) [1 – (2/A)] A_3 = 1/(MmAB) a_w = water activity A, A_1, A_2, A_3, B, C = constants EMC = equilibrium moisture content Mm = GAB monolayer
BET [24]	$\frac{a_w}{EMC(1-a_w)} = \frac{1}{M_0C} + \frac{C-1}{M_0C}a_w$	M_o = BET monolayer value C = constant for a given water sorbent system a_w = water activity EMC = equilibrium moisture content

Table 1.
Sorption isotherms models equations.

Where:
w_f = final weight of sample.
w_i = initial weight of sample.
$\%H_2O$ = initial moisture content.

2.2.4 Sorption equations

To smooth the model of adsorption and/or desorption curves, empirical models that can describe the relationship between water equilibrium content, relative humidity, and temperature exist. Indeed, both the GAB (Guggenheim–Anderson–De Boer) and BET (Brunauer, Emmett, and Teller) models are the best formulas used to determine the monolayer of a food, applicable for water activities between 0.05 and 0.95. The GAB monolayer value can be estimated using linear or non-linear methods. The difference in percentage in EMC between the duplicate samples was on average < 5% when the average of the two values was taken. The BET value is found by plotting $a_w/(1-a_w)EMC$ versus a_w and using the intercept to cover the monolayer. Both GAB and BET sorption equations were used to analyze the sorption isotherm data (**Table 1**).

2.2.5 Statistical analysis

Data were obtained in triplicate. One-way analysis of variance (ANOVA) was performed, and significant differences in mean values were evaluated by Tukey HSD multiple range test at ($P < 0.05$) using SPSS version 17.0 (SPSS, Chicago, IL, USA). Microsoft Excel 2013 was used to carry out the regression analysis of water activity as a function of increasing relative humidity and constant temperature of cowpea varieties flour.

3. Results and discussion

The cowpea grains processing technologies used by the local population of the sub-Saharan region are less efficient. The use of heterogeneous and unsuitable packaging, as well as traditional drying and preservation methods, constitutes constraints for this activity to flourish [16, 25]. Indeed, drying influences, in particular, the water content and consequently some physicochemical, rheological, organoleptic, and functional characteristics of the food product. The moisture content of the three cowpea varieties flour samples varies from 07.52 to 08.28%, as presented in **Table 2**. The oil absorption capacities of cowpea varieties flour are found to be 1.06 ± 0.01, 1.08 ± 0.02, and 1.09 ± 0.01, respectively, for CS133, CS032, and control sample mL/g. The water-holding capacity and the swelling index were revealed to be significantly dependent ($P < 0.05$) from one to another. Thus, the results showed that the water absorption capacity ranged, respectively, for CS133, CS032, and control from 1.01 ± 0.17, 0.83 ± 0.10, and 0.69 ± 0.11 g/g. The swelling index data varied to 4.10 ± 0.29, 3.78 ± 0.02, and 3.67 ± 0.24 g/cm^3 for CS133, CS032, and control (**Table 2**).

This can be explained that flour which absorbs less water absorbs more oil and vice versa. These results corroborate with those of Naiker [6]. Moreover, a high-water content of flour could promote chemical and enzymatic reactions, the development of microorganisms leading to the deterioration of the product quality. In addition, the water content of food products plays a key role in their preservation [4, 26]. Thus, it appears essential to determine the minimum water content that can promote the preservation of cowpea varieties flour. The ability of a food to absorb more water can also be attributed to its content of protein and carbohydrates with free hydrophilic residues. The result shows that CS133 flour swells more than other varieties; this could be explained by the fact that flour that absorbs more water content swells better indeed. Flour with a good swelling index could be of good quality and used in pastry [11, 27]. Furthermore, a quality flour stability depends also on its drying conditions and preservation [17, 28].

Isotherms are particularly important for determining the minimum water content of a food product. Two mathematical models (GAB and BET) were used to obtain equilibrium moisture content (EMC). The data of EMC of cowpea varieties flour obtained experimentally at the temperatures of 30, 40, and 50°C, and for each of the relative humidity or the water, activities are presented in **Table 3**.

However, the average EMC values are used for the representations of adsorption isotherms by samples of cowpea varieties flours. The GAB model happened to be the most adequate model for the smoothing of the adsorption isotherms of cowpea varieties flour in this study, contrary to the findings where BET model was best smooth of sorption isotherms in the study on the dehydrated beef made in Nigeria [12]. The different adsorption curves of the cowpea varieties flour related to temperature, and different varieties are shown in **Figure 1**.

Practically, the hygroscopic equilibrium was observed around 2 weeks later, and it was found that the greater the water activity (a_w) is maintained in the medium, the easier it is for the water content to be determined. Therefore, this shows the influences of relative humidity and temperature on the drying of each of the cowpea

Cowpea varieties	Moisture content (%)	Oil adsorption capacity (mL/g)	Water adsorption capacity (g/g)	Swelling index (g/cm^3)
CS133	9.92 ± 0.03a	1.06 ± 0.01a	1.01 ± 0.17a	4.10 ± 0.29a
CS032	7.96 ± 0.05a	1.08 ± 0.02a	0.83 ± 0.10a	3.78 ± 0.02b
Control	6.97 ± 0.13a	1.09 ± 0.01a	0.69 ± 0.11a	3.67 ± 0.24c

Table 2.
Physico-functional properties of cowpea varieties flour.

Temperature (°C)	Relative humidity (%)	a_w	EMC		
			CS133	CS032	Control
30	10	0.1	1.09	1.54	1.92
	30	0.3	1.74	1.91	2.09
	50	0.5	2.27	2.59	3.58
	70	0.7	3.84	4.58	4.44
	80	0.8	4.15	5.1	5.43
	90	0.9	9.28	11.28	10.51
40	10	0.1	1.26	1.42	1.21
	30	0.3	1.75	2.27	2.35
	50	0.5	3.19	2.37	2.46
	70	0.7	4.62	4.12	4.09
	80	0.8	9.11	4.89	4.67
	90	0.9	10.25	11.4	10.69
50	10	0.1	1.29	1.34	1.37
	30	0.3	2.2	2.42	2.24
	50	0.5	3.18	4.45	3.28
	70	0.7	3.35	5.97	4.19
	80	0.8	4.29	11.3	5.33
	90	0.9	11.8	21.26	12.75

Table 3.
Equilibrium moisture content (EMC) of cowpea varieties flour as a function of water activity (a_w).

variety flour. Likewise, previous research on food products has demonstrated that the sorption isotherm curves are represented in a sigmoidal shape [12, 28]. In fact, it can be noted that only the adsorption isotherm curve of the sample CS032 at 50°C showed a strong correlation different from other samples at different temperatures (**Figure 2**) with $R^2 = 0.9274$ and $R^2 = 0.8226$ respectively for GAB and BET models.

It can be seen that the more the temperature increases, the less noticeable the difference in the drying of the inter-varietal cowpea flour is. In this line, the sorption isotherms are of type II, that is the sigmoidal sorption isotherms, in which the curves are concave upwards, taking into account the existence of multilayers at the internal surface of cowpea flour material. Generally, the increase in temperature induces the decrease in EMC, leading to an increase in a_w for constant water content. For lower a_w, the EMC is also low, compared to levels when the a_w approaches one (1) [29, 30]. Knowingly, the EMC decreases with increasing temperature and for the same value of relative humidity (RH) leads to the endothermic reaction [31]. It was observed for the samples CS133 and control for the temperatures of 30 to 50°C that the water content decreases when the temperature increases within the interval values of a_w between 0.37 and 0.78%, of which most of the curves are found around 0.8 a_w. Indeed, this decrease can be caused by the increase in the heat of absorption in the case of high temperatures, which makes it possible to reduce the EMC [32, 33]. Benseddik et al. and Ferradji and Matallah [26, 34] stipulated that they observed changes in sorption

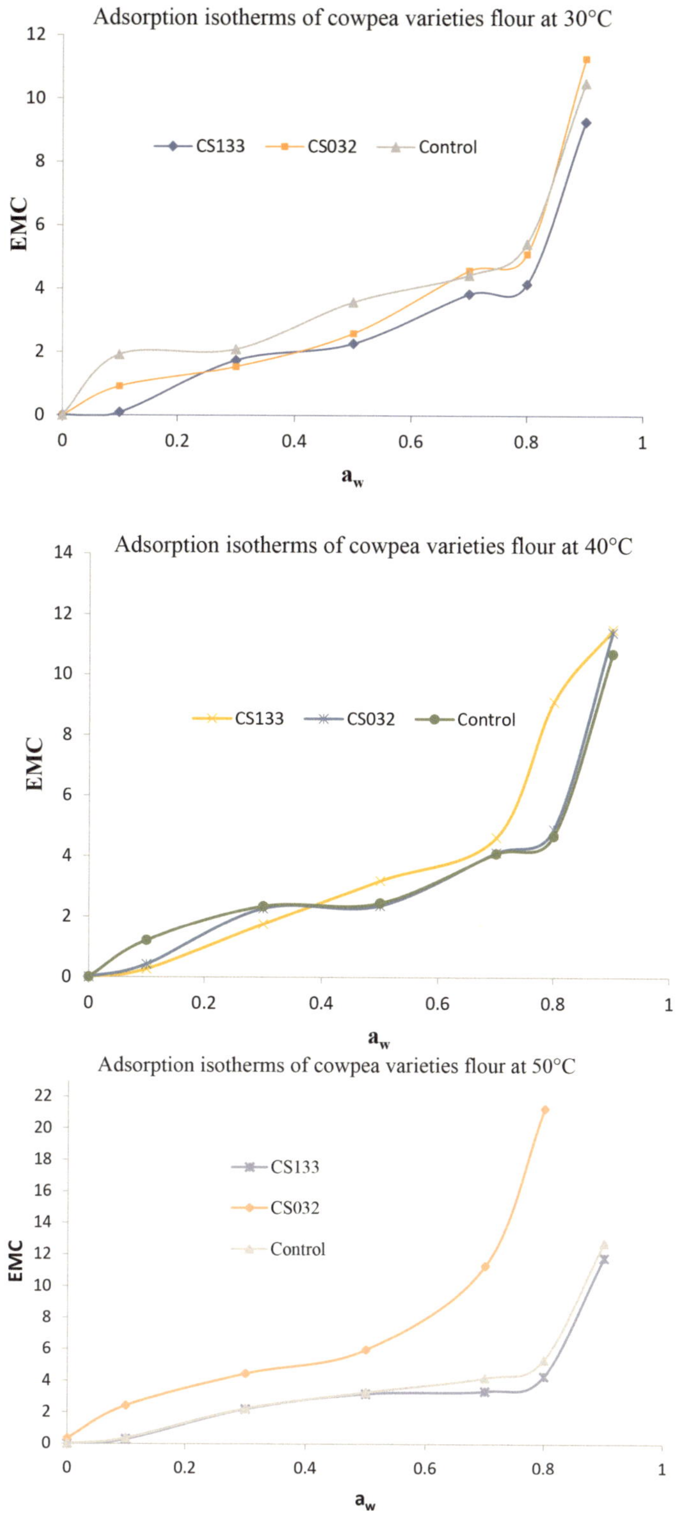

Figure 1.
Adsorption isotherms of cowpea varieties flour by temperatures 30(a), 40(b) and 50(c) °C).

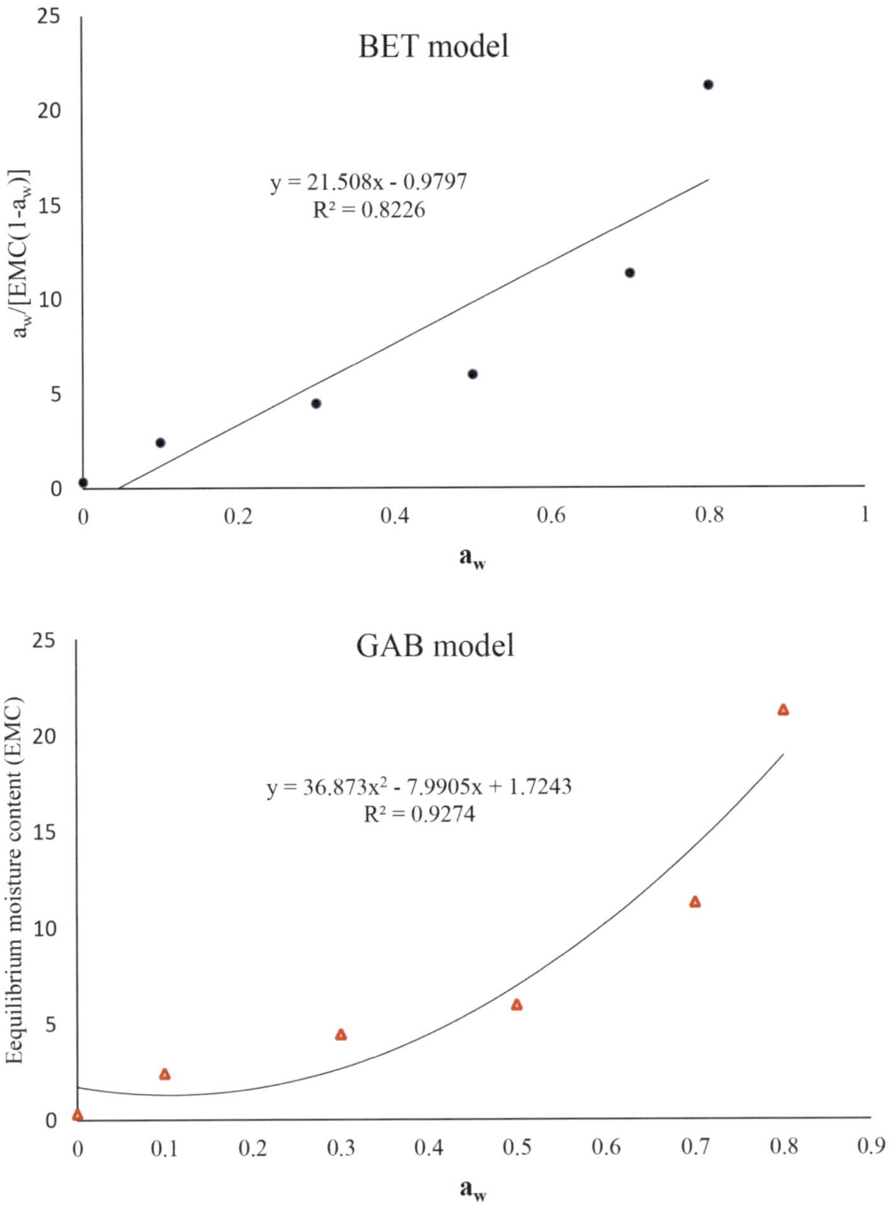

Figure 2.
Adsorption isotherms curves of CS032 flour at 50°C smoothed by BET and GAB.

isotherms with an increase in temperature is relative to the composition of the products in starch content and its solubility with an increase in thermal agitation within the samples [35]. Though, at high temperatures, the state of excitation is stronger and favors the reduction of the forces of attraction of molecules among them (**Figure 3**).

Moreover, the processing of flour regardless of the variety is considered to depend on its end-use; as result, in certain products, the more the sugars dissolve, the more mobility and the availability of water are reduced. In addition, the ability of water adsorption can also be attributed to its content of protein. Moreover, for products

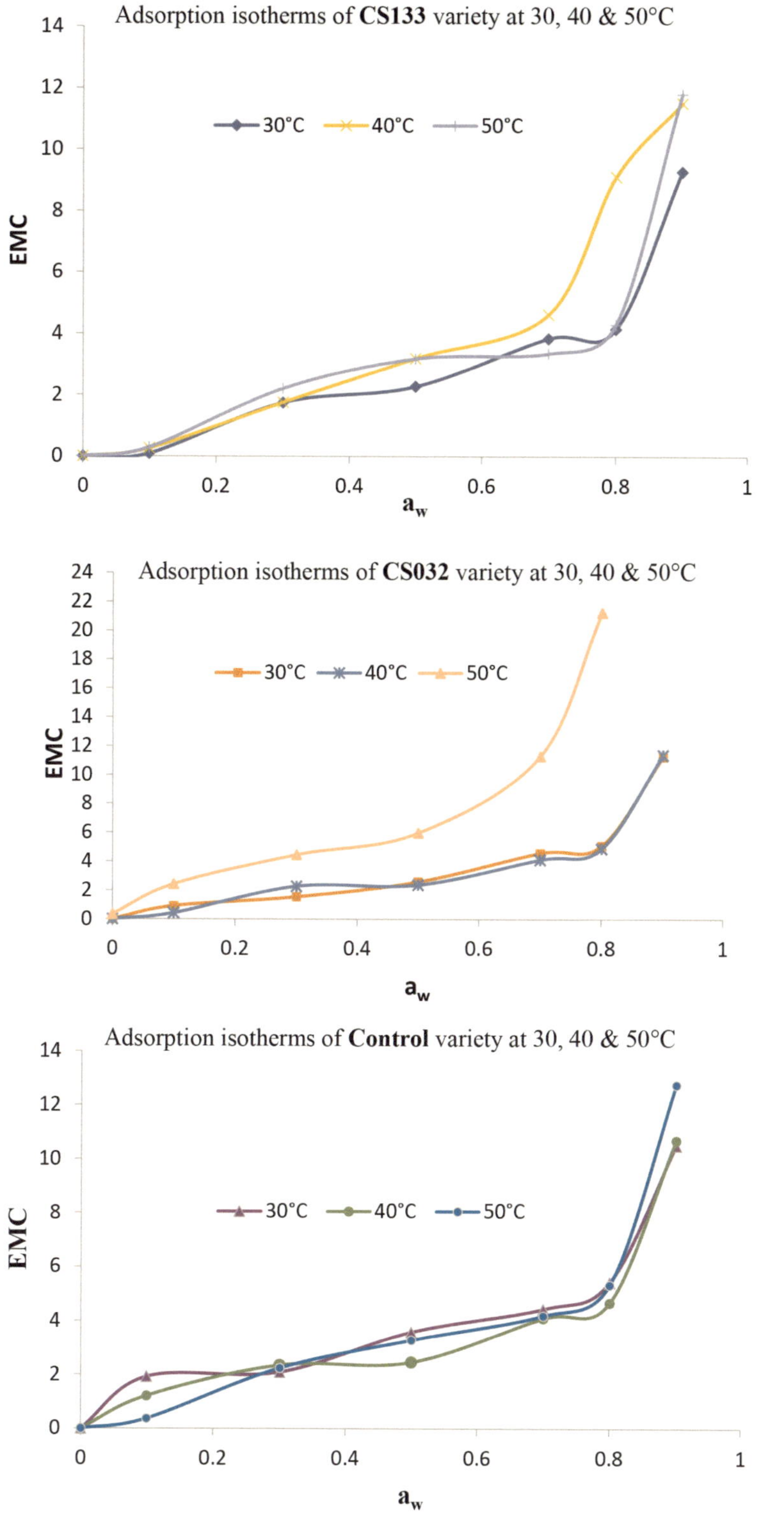

Figure 3.
Adsorption isotherms with temperature differences on the cowpea varieties flour (CS133, CS032 and control).

with low sugar content such as flour, the curves do not intersect, likewise, the work by Koko et al. [4]. The flour of cowpea varieties is not an exception of this trend, whereas the physicochemical properties further affect their drying conditions.

4. Conclusion

It can be concluded that the flour of cowpea varieties absorbs differently the water and oil, and the swelling index significantly. It was a fact that flour, which absorbs more water content, swells better, as far as the sorption isotherms of cowpea varieties flour were carried out at different temperatures range of 30, 40, and 50°C. The data showed a sigmoidal shape, characteristic of type II isotherms, whether determined by the GAB or BET models. It was found that the GAB model allows the most important relative squared errors. Thus, the GAB model was the most adequate model for the moisture absorption isotherms on the flours of the three varieties of cowpea studied. For lower aw, the EMC was also low, compared to levels when the a_w approaches one (1). It was noted that the more the surrounding temperature increases, the more the water content of the product decreases. A significant linear interaction was revealed to better describe the variation within the EMC range considered in this work. Thus, depending on the rate of swelling of these varieties, it can be deduced that the CS032 cowpea variety flour seemed to behave better than its counterpart varieties. More such findings are needed to favor these legume development programs for expanding preferable varieties for targeted applications.

Acknowledgements

This work was supported by the CowpeaSquare II project MCKNIGHT ID: 15-114, and the author is grateful to Mr. Yacouba Issoufou Maaroufi for technical assistance and project coordination team.

Author contributions

Issoufou AMADOU designed, supervised lab technicians for the experiments, wrote and proofread the article.

Funding

This research was funded by the CowpeaSquare II project MCKNIGHT, grant number 15-114.

Conflicts of interest

The author declares no conflict of interest.

Abbreviations

a_w	Water activity
BET	Brunauer, Emmett, and Teller
CS133 and CS032	Names of cowpea varieties for CowpeaSquare
EMC	Equilibrium moisture content
GAB	Guggenheim - Anderson - De Boer
UDDM	Unversité Dan Dicko Dankoulodo de Maradi

Author details

Issoufou Amadou
Laboratory of Food Science and Technology, Faculty of Agronomy and Environmental Sciences, Dan Dicko Dankoulodo University of Maradi, Niger

*Address all correspondence to: issoufsara@gmail.com

References

[1] Raina A, Laskar RA, Tantray YR, Khursheed S, Wani MR, Khan S. Characterization of induced high yielding cowpea mutant lines using physiological, biochemical and molecular markers. Scientific Reports. 2020;**10**(1):1-22. DOI: 10.1038/s41598-020-60601-6

[2] Boukar O, Abberton M, Oyatomi O, Togola A, Tripathi L, Fatokun C. Introgression breeding in cowpea [*Vigna unguiculata* (L.) walp.]. Frontiers in Plant Science. 2020;**1-11**(11):56745. DOI: 10.3389/fpls.2020.567425

[3] Lanza MGDB, Silva VM, Montanha GS, Lavres J, de Carvalho HWP, Dos Reis AR. Assessment of selenium spatial distribution using μ-XFR in cowpea (*Vigna unguiculata* (L.) Walp.) plants: Integration of physiological and biochemical responses. Ecotoxicology and Environmental Safety. 2021;**207**: 111216. DOI: 10.1016/j.ecoenv.2020.111216

[4] Koko CA, Diomande M, Kouame BK, Assidjo EN. Détermination expérimentale et modélisation des isothermes d'adsorption d'eau des amandes d'*Irvingia gabonensis* de la région du Haut-Sassandra (Côte d'Ivoire). IOSR Journal of Environmental Science, Toxicology and Food Technology. 2018;**12**(2):50-66. DOI: 10.9790/2402-1202025066

[5] Harou A, Hamidou F, Bakasso Y. Performances morpho-physiologiques et agronomiques du niébé [*Vigna unguiculata* (L.) Walpers] en conditions du déficit hydrique. Journal of Applied Bioscience. 2018;**128**:12874-12882. DOI: 10.4314/jab.v128i1.1

[6] Naiker TS, Gerrano A, Mellem J. Physicochemical properties of flour produced from different cowpea (*Vigna unguiculata*) cultivars of Southern African origin. Journal of Food Science and Technology. 2019;**56**(3):1541-1550. DOI: 10.1007/s13197-019-03649-1

[7] Appiah F, Asibuo JY, Kumah P. Physicochemical and functional properties of bean flours of three cowpea (*Vigna unguiculata* L. Walp) varieties in Ghana. African Journal of Food Science. 2011;**5**(2):100-104. DOI: 10.17660/actahortic.2011.911.51

[8] Kebede E, Bekeko Z. Expounding the production and importance of cowpea (*Vigna unguiculata* (L.) Walp.) in Ethiopia. Cogent Food & Agriculture. 2020;**6**(1):1769805. DOI: 10.1080/23311932.2020.1769805

[9] da Silva AC, da Costa Santos D, Junior DLT, da Silva PB, dos Santos RC, Siviero A. Cowpea: A strategic legume species for food security and health. In: Legume Seed Nutraceutical Research. London, UK: IntechOpen; 2018. DOI: 10.5772/intechopen.79006

[10] Hawa LC, Ubaidillah U, Damayanti R, Hendrawan Y. Moisture sorption isotherms of modified cassava flour during drying and storage. Heat and Mass Transfer. 2020;**56**(8):2389-2396. DOI: 10.1007/s00231-020-02866-1

[11] Luthra K, Shafiekhani S, Sadaka SS, Atungulu GG. Determination of moisture sorption isotherms of rice and husk flour composites. Applied Engineering in Agriculture. 2020;**36**(6): 859-867. DOI: 10.13031/aea.13822

[12] Amadou I, Diadie HO, Gbadamosi OS, Akanbi CT. Characterization and sorption isotherm of dehydrated beef made in Nigeria. Cogent Food & Agriculture. 2019;**5**(1):

1710440. DOI: 10.1080/23311932.2019.1710440

[13] Muangrat R, Nuankham C. Moisture sorption isotherm and changes in physico-mechanical properties of films produced from waste flour and their application on preservation quality of fresh strawberry. Food Science & Nutrition. 2018;**6**(3):585-593. DOI: 10.1002/fsn3.589

[14] Luo Y, Liu Q, Liu J, Liu X, Zhao S, Hu Q, et al. Effect of starch multi-scale structure alteration on japonica rice flour functionality under infrared radiation drying and storage. Lebensmittel-Wissenschaft & Technologie, Journal of Food Science and Technology. 2021;**143**: 111126. DOI: 10.1016/j.lwt.2021.111126

[15] Coradi PC, Müller A, Souza GA, Steinhaus JI, Wagner R. Quality of soybean cultivars in the drying and storage processes in real scale and experimental. Journal of Food Process Engineering. 2020;**43**(7):e13418. DOI: 10.1111/jfpe.13418

[16] Yadav N, Kaur D, Malaviya R, Singh M, Fatima M, Singh L. Effect of thermal and non-thermal processing on antioxidant potential of cowpea seeds. International Journal of Food Propertie. 2018;**21**(1):437-451. DOI: 10.1080/10942912.2018.1431659

[17] Ngoma TN, Chimimba UK, Mwangwela AM, Thakwalakwa C, Maleta KM, Manary MJ, et al. Effect of cowpea flour processing on the chemical properties and acceptability of a novel cowpea blended maize porridge. PLoS One. 2018;**13**(7):e0200418. DOI: 10.1371/journal.pone.0200418

[18] Oyeyinka SA, Kayitesi E, Adebo OA, Oyedeji AB, Ogundele OM, Obilana AO, et al. A review on the physicochemical properties and potential food applications of cowpea (*Vigna unguiculata*) starch. International Journal of Food Science and Technology. 2021; **56**(1):52-60. DOI: 10.1111/ijfs.14604/v1/review2

[19] AOAC. Association of Official Analytical Chemists. Methods 932.06, 925.09, 985.29, 923.03: Official methods of analysis of the AOAC. Arlington, VA: AOAC; 1990. In Association of official analytical chemists (15th)

[20] Sofi BA, Wani IA, Masoodi FA, Saba I, Muzaffar S. Effect of gamma irradiation on physicochemical properties of broad bean (*Vicia faba* L.) starch. Lebensmittel-Wissenschaft & Technologie. 2013;**5**:63-72. DOI: 10.1016/j.lwt.2013.05.021

[21] Matz SA. The Physical and Technology of Cereals as Food and Feed. 2nd ed. New York: Springer; 1991. p. 752

[22] Labuza TP. In: St Paul MN, editor. Practical Aspect of Moisture Sorption Isotherms, Measurement and Use. St. Paul, Minnesota: American Association of Cereal Chemistry, Press; 1984. ISBN: 1891127187 9781891127182

[23] Maroulis ZB, Tsami E, Marinos-Kouris D, Saravacos GD. Application of the GAB model to the moisture sorption isotherms for dried fruits. Journal of Food Engineering. 1988;**7**(1):63-78. DOI: 10.1016/0260-8774(88)90069-6

[24] Ayeranci E, Ayranci G, Dogantan Z. Moisture sorption isotherms of dried apricots, fig and raisin at 20°C and 30°C. Journal of Food Science. 1990;**55**: 1591-1593. DOI: 10.1111/j.1365-2621.1990.tb03577.x

[25] Bighaghire R, Okidi L, Muggaga C, Ongeng D. Traditional vegetable preservation technologies practiced in Acholi subregion of Uganda improves

mineral bioavailability but impacts negatively on the contribution of vegetables to household needs for micronutrients. Food Science & Nutrition. 2021;**9**(2):589-604. DOI: 10.1002/fsn3.1931

[26] Ferradji A, Matallah MAA. Sorption isotherms and isosteric heats for Algerian dates Deglet Nour. American Journal of Food Technology. 2012;**7**(6): 352-362. DOI: 10.3923/ajft.2012.352.362

[27] Huang S, Bohrer BM. The effect of breadfruit (*Artocarpus Altilis*) flour on textural properties of comminuted beef compared with other flour sources. Meat and Muscle Biology. 2018;**2**(2):73-73. DOI: 10.22175/rmc2018.063

[28] Brett B, Figueroa M, Sandoval AJ, Barreiro JA, Müller AJ. Moisture sorption characteristics of starchy products: Oat flour and rice flour. Food Biophysics. 2009;**4**(3):151-157. DOI: 10.1007/s11483-009-9112-0

[29] Bergman R. Drying and control of moisture content and dimensional changes. In: Chapter 13 in FPL-GTR-282. Madison, WI, USA: USDA Forest Service, Forest Products Laboratory. 13–1. 2021

[30] Mallek-Ayadi S, Bahloul N, Kechaou N. Mathematical modeling of water sorption isotherms and thermodynamic properties of *Cucumis melo* L. seeds. Lebensmittel-Wissenschaft & Technologie. 2020;**131**: 109727. DOI: 10.1016/j.lwt.2020.109727

[31] van't Hag L, Danthe J, Handschin S, Mutuli GP, Mbuge D, Mezzenga R. Drying of African leafy vegetables for their effective preservation: The difference in moisture sorption isotherms explained by their microstructure. Food & Function. 2020; **11**(1):955-964. DOI: 10.1039/c9fo01175g

[32] Lopez-Quiroga E, Prosapio V, Fryer PJ, Norton IT, Bakalis S. Model discrimination for drying and rehydration kinetics of freeze-dried tomatoes. Journal of Food Process Engineering. 2020;**43**(5):e13192. DOI: 10.1111/jfpe.13192

[33] Zambrano MV, Dutta B, Mercer DG, MacLean HL, Touchie MF. Assessment of moisture content measurement methods of dried food products in small-scale operations in developing countries: A review. Trends in Food Science and Technology. 2019;**88**:484-496. DOI: 10.1016/j.tifs.2019.04.006

[34] Benseddik A, Azzi A, Zidoune MN, Allaf K. Mathematical empirical models of thin-layer airflow drying kinetics of pumpkin slice. Engineering in Agriculture, Environment and Food. 2018;**11**(4):220-231. DOI: 10.1016/j.eaef.2018.07.003

[35] Habibiasr M, Mokhtar MN, Ibrahim MN, Md Yunos KF, Ibrahim NA. Study on the effects of physical properties of Tenera palm kernel during drying and its moisture sorption isotherms. PRO. 2020;**8**(12): 1658. DOI: 10.3390/pr8121658

Section 2

Soprtion in Various Substrates

Chapter 4

Transition Metals-Based Metal-Organic Frameworks, Synthesis, and Environmental Applications

Lidia E. Chiñas-Rojas, Guadalupe Vivar-Vera, Yafeth F. Cruz-Martínez, Seth Limón Colohua, José María Rivera and Eric Houbron

Abstract

This work illustrates examples of metal-organic frameworks (MOFs) derived from transition metals and their environmental applications in areas of catalysis, sorption, and hydrogen evolution. Explanation of some of the techniques employed for their synthesis has been discussed. On the other hand, the advantages of the use of hybrid materials such as the metal-organic frameworks are exposed in this book as well a detailed description of the different linkers and metals used for the synthesis of this kind of porous materials going through the methodologies and techniques utilized by different authors to obtain good-quality crystalline applicable materials. Adjustments of linker geometry, length, ratio, and the functional group can tune the size, shape, and internal surface property of an MOF for a targeted application. The uses of MOFs are exploring new different areas of chemistry such as catalysis, adsorption, carrier systems, hydrogen evolution, photocatalysis, and more. Different examples of MOFs from Scandium to Zinc are well described in this book, and finally, a brief description of some common environmental applications such as metals and azo dyes sorption, hydrogen evolution, and catalyst in the transesterification process of vegetable oils to produce biodiesel is explored and commented.

Keywords: coordination, polymers, solvothermal, sorption, X-diffraction, metal-organic frameworks

1. Introduction

Metal-organic frameworks (MOFs) have been widely reported in the literature during the last two decades, and the number of articles published is exponentially increasing due to the opportunity to obtain a great diversity of novel crystalline porous materials with different topologies and most important with a countless number of applications in different fields of chemistry such as catalysis, photocatalysis, gas storage, separation, sorption, hydrogen evolution, and more [1–9].

In general, crystalline porous materials can have several terminologies, for example, metal-organic material (MOM), coordination polymer (CP), coordination network (CN), porous coordination polymer (PCP), porous coordination network (PCN), microporous coordination polymer (MCP), and metal-organic coordination network (MOCN), which are habitually used by different scientists to designate, at least some, MOFs. All these subclasses of coordination compounds have very similar structural features with slight differences and therefore confer an inevitable overlap in their properties. For example, coordination polymer (CP) is a common term that has occasionally been used as an alternative word for MOF. The term "coordination polymer" implies that coordination compounds that constitute one-, two-, or three-dimensional (1D, 2D, or 3D) polymeric structures via linking of the metal ions by bridging ligands are CPs [1]. The synthesis of MOFs requires two principal components, the ligands commonly called linkers, and the metals in the form of different salts such as nitrates, chlorates, perchlorates, or sulfates. The ligands are usually any atom or any organic molecule, the latter possesses in its structure different functional groups as carboxylic acid, amine, pyridine group, among others, with the ability to donate at list a lone electron pair to the metal, commonly known as Lewis's base. The pore size depends mostly on the length of the ligand. On the other hand, the second component corresponds to the metal, which may vary from the s-block, p-block, transition metals, or even rare earth metals. The obtention of MOFs in 1D, 2D, and 3D dimensions depends directly on the metal employed for the synthesis, which has to do with its oxidation state and the types of geometries that the metal can acquire. And finally, a very important issue to consider is the technique used to obtain the MOFs that involves several parameters such as solvent or solvent mixture, temperature, pressure, pH value among others, all these variables together may or may not give rise to the desired materials. Different synthesis methods to obtain MOFs have been applied in the last 20 years in addition to the room temperature or slow evaporation synthesis, which are conventional electric (CE) heating, microwave (MW) heating, electrochemistry (EC), mechanochemistry (MC), ultrasonic (US) methods, and a combination of the previous techniques before mentioned. Conventional step-by-step methods, as well as high-throughput methods, have been employed in some of the studies. There have been various studies regarding the morphology of the MOF products, in addition to the crystal size or shape, thin films, membranes, and various other shapes made of MOFs have been reported, which require the application of different synthesis methods [2].

2. Linkers

The geometry and connectivity of a linker dictate the structure of the resulting MOF, which are broadly classified into neutral (MOFs) and ionic (iMOFs), based on the charge of the framework backbone. Ionic MOFs (iMOFs) are subdivided into cationic (iMOF-C) and anionic (iMOF-A). Azolate based linkers may give neutral as well as ionic framework depending on the connectivity of the linker, metal coordination environment, and coordination geometry. **Figure 1** shows the type of MOF obtained according to the N-donor linker base used [3].

Many of the systems reported from the use of azolate-based linkers are neutral in nature, wherein the charge balance is either via anionic donor groups of azolates or the presence of coordinating anion used in the synthesis. For instance, all the ZIF-series MOFs, which are constructed from imidazolates, are electrically neutral. Likewise,

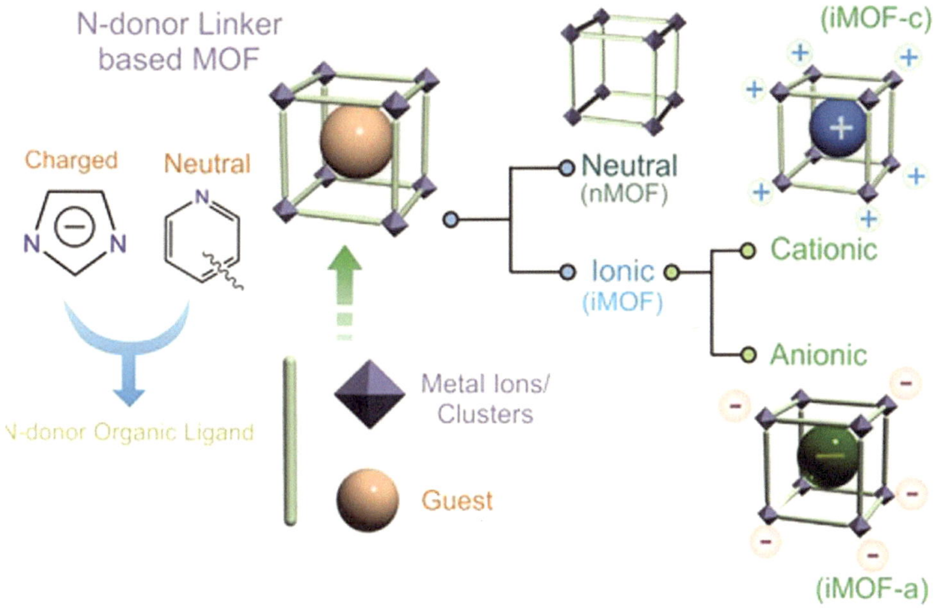

Figure 1.
Classification of MOFs constructed from N-donor linkers [3].

pyrazolate and triazolate-based systems have been found to self-assemble with metal ions/clusters into neutral MOFs. A good example of a nitrogen-containing linker is the 4,4′-bipyridine, which is an aromatic organic compound with nitrogen hetero atoms located in the para positions of the aromatic rings. Nitrogen atoms possess an electron lone pair located in the plane of the carbon atoms of the ring, such arrangement allows a very specific coordination binding to the metal. **Figure 2** shows different N-donor linkers used in the synthesis of coordination polymeric materials [3].

Typical N-donor linkers can have 1, 2, 3, 4 or more nitrogen atoms. It's very important to select the appropriate linker to obtain the desired pore size, i.e., the linker 4,4′-bipyridine, which was reported by Wang and coworkers, describes a MOF synthesized through hydrothermal conditions and using two different linkers [4]. As could be measured from the .cif file, the distance between the Cd atoms is in the range of 11.59–7.50 Å. Also, the rings of the N-ligands are almost planar showing π-π interactions. Due to the electron-rich cloud inside the pore, small cations, or even small molecules with a deficiency of electrons, could interact with each other to form stable systems. This analysis can be deduced from the packing material obtained by X-ray diffraction. **Figure 3** shows a small portion of the material, which is composed of two different organic linkers where the cadmium atoms are hexacoordinated arranged in an octahedral fashion [4].

The exchange of the 4,4′-bipyridine linker for a small one like the pyrazine linker leads to a decrease in the distance between the metal atoms and hence the pore size is reduced. Arenzano and coworkers reported the pillared MOF shown in **Figure 4**, which is composed of two different organic linkers, pyrazine and 2-amine isophthalic acid, which possess free amino groups in the final structure that can interact with other atoms or molecules by hydrogen bonding or even donating the electron pair to another metal atom to form a coordination bond. In the same way as the material reported by Wang, this keeps the π-π interactions between the pyrazine rings.

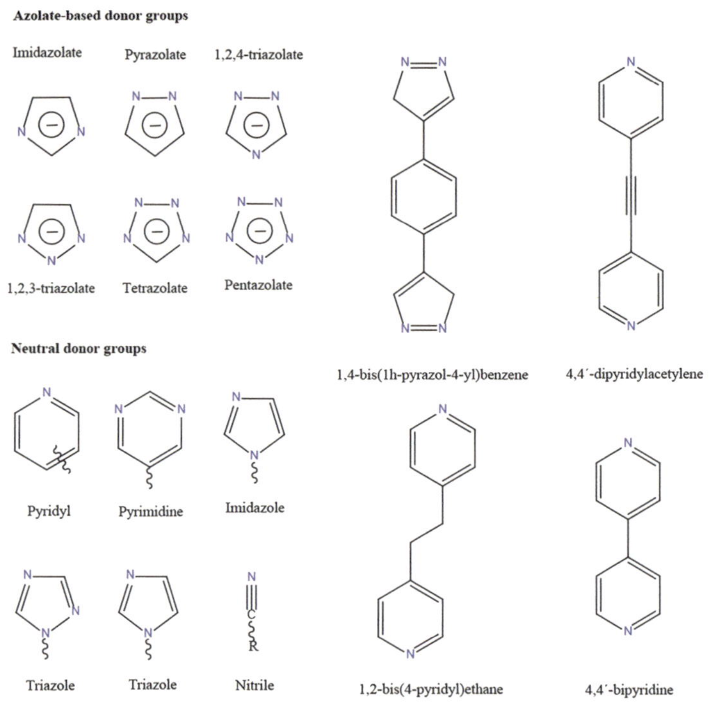

Figure 2.
Typical donor groups in N-donor linkers.

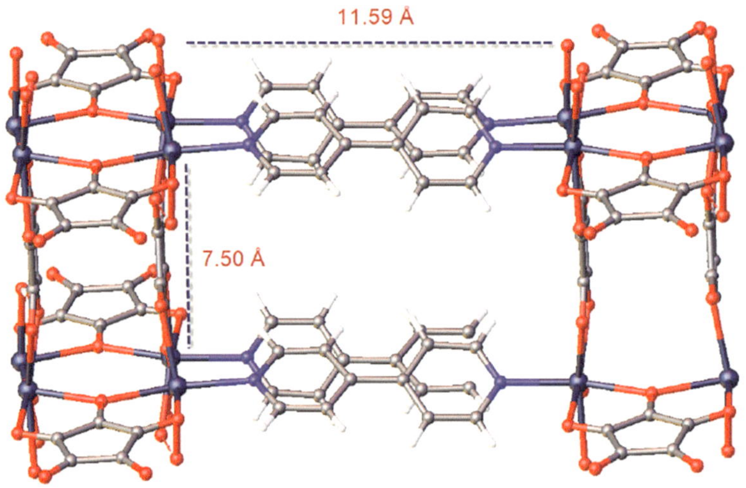

Figure 3.
Pore size and distances between Cd atoms in the MOF described by Wang.

The distances between cobalt atoms are in the range from 7.34 Å to 7.14 Å. The material that possesses free functional NH_2 groups able to form hydrogen bonding interactions can be used as a carrier system [5].

Carboxylate linkers possess two oxygen atoms, which contain four-electron lone pairs ready to be donated to a metal to form a coordination bond; for this reason, the carboxylate ligand, $[RCO_2]^-$, is one of the most versatile and hence also one of the

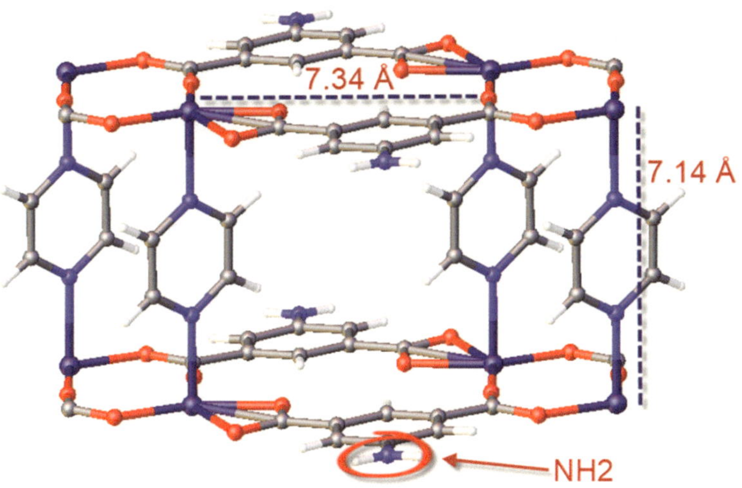

Figure 4.
Distances between cobalt atoms and free amino groups present in MOF.

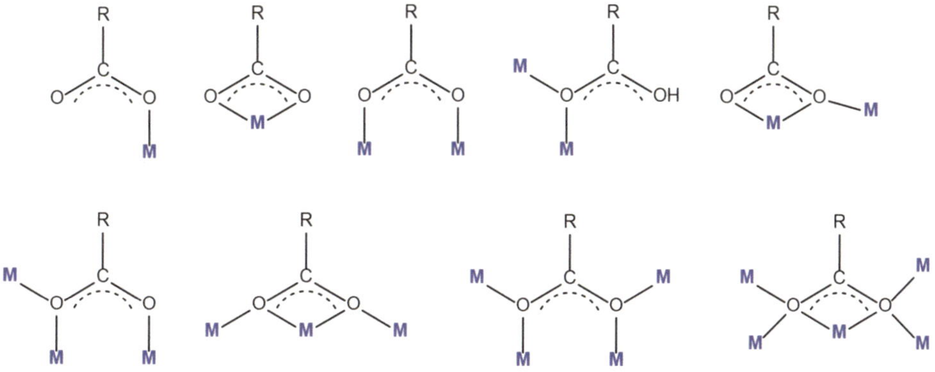

Figure 5.
Carboxylate coordination modes.

most abundant ligands in coordination chemistry. A great difference between the N-ligands and carboxylate ligands is the coordination modes. Although simple coordination modes are exhibited by the carboxylate and multicarboxylate family of ligands, their utility has been immense, **Figure 5** [6].

Designing the network topology in MOFs can be done within two main routes, changing the cluster connectivity or altering the linker's topological geometry. In the same way as the N-linkers, the carboxylate length may vary the pore size in the final structure. Linkers can be functionalized, that is, functional groups are added in different sites of the benzene ring to confer certain desired properties in the final material [6–10]. **Figure 6** shows different carboxylate linkers, which differ mainly in length, generally used in the synthesis of coordination polymers. Metal clusters adopt established geometries that cannot be easily modified to change the connectivity. The creation and modification of pore space with optimized size, functionality, and diversity can be precisely tuned at the molecular level by rationally designing building blocks and synthetic procedures. Therefore, logical designing of linker geometry is essential for discovering specified topologies as an ideal platform for designing MOFs.

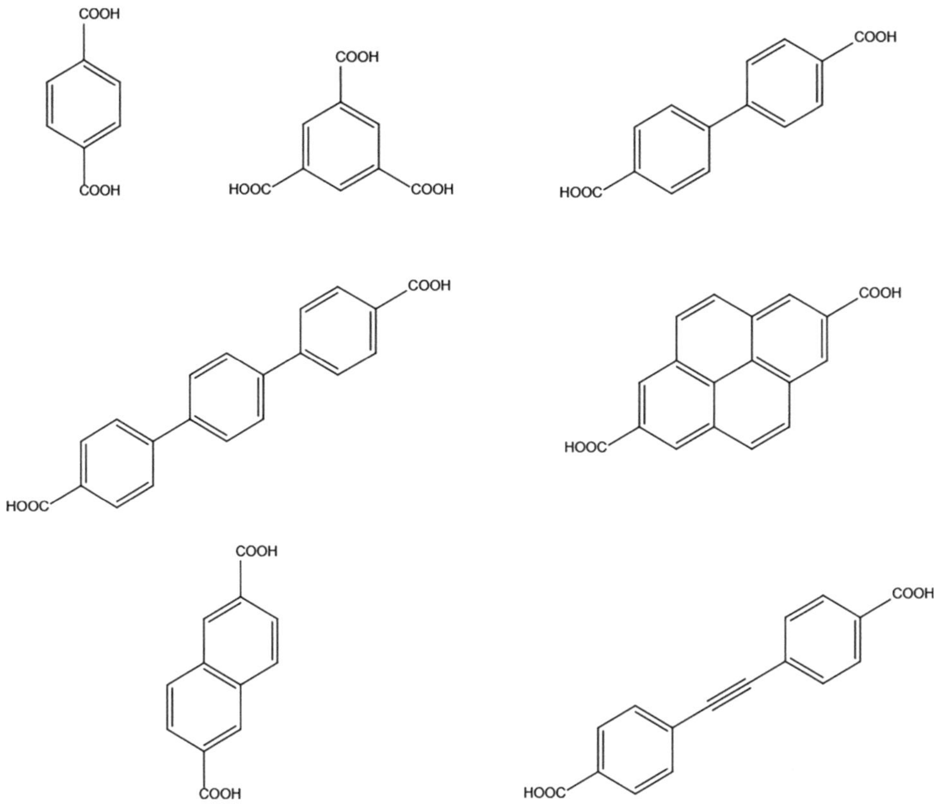

Figure 6.
Linkers based on the carboxylate functional group to synthesize MOFs.

3. S-block metals-based MOFs

MOFs made up of s-block main group elements, alkali and alkaline earth metals, have been less considered in both the fundamental and applied chemistry because of their low stability and most importantly, preconceived chemical features Despite the limited structures and difficulties in the structural formation of s-block MOFs originating from the unpredicted coordination behavior of metals in group 1A and 2A, s-block MOFs adopt many distinctive, appealing, and intriguing features that are suited to many applications. S-block MOFs inherit the characteristics of s-block main group elements that are (1) naturally high abundance, making them inexpensive; (2) less toxicity, unlike the other MOFs based on transition metals, thus expanding their applicability into various fields, especially in biological processes; and (3) low density, one of the essential factors for gas sorption applications [11].

4. Transition metal-based MOFs

4.1 Scandium MOFs

Scandium-based MOFs are stable and widely reported in the literature, reports on highly selective CO_2 capture by small pore scandium-based MOF, hydrogen storage,

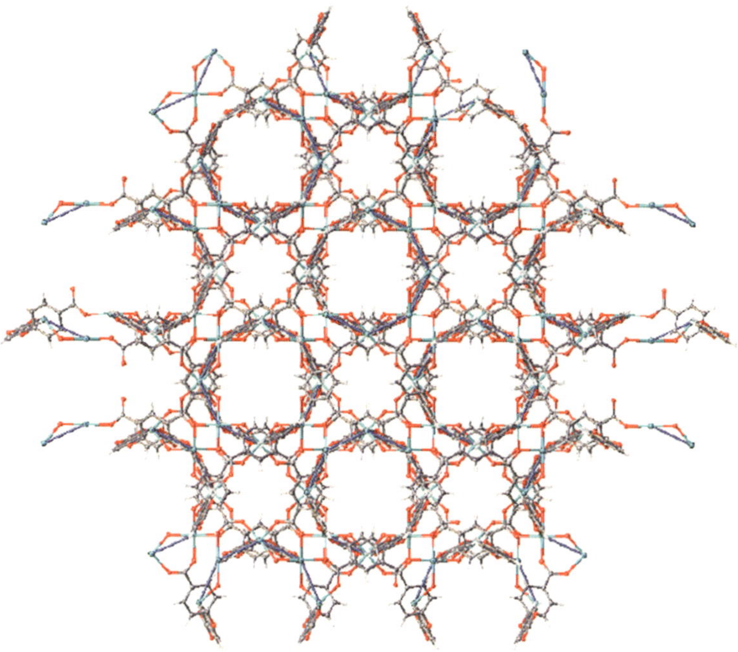

Figure 7.
Types of channels with electronic charge distribution observed in red color in MOF Sc-CAU-21.

catalysis, sorption, and more [12–15]. Recently, Stock and coworkers described a Scandium MOF, which was synthesized under solvothermal reaction conditions using 4,4′-oxidibenzoic. The crystal structure of Sc-CAU-21 was determined from single crystal X-ray diffraction data and showed two different types and sizes of channels. Cadmium atoms present an octahedral geometry with bond distances in the range of 2.059–2.083 Å. **Figure** 7. The pore size can give us an idea of the molecules, which could be absorbed in the Sc-CAU-21 [16].

The IBUs (Inorganic Building Units) are connected by 4,4′-oxidibenzoate linker molecules to form a 3Dframework, which is isostructural to the nonporous Al-CAU-21. Doping of Sc-CAU-21 was carried out to tune the luminescence properties where Sc-CAU-21 showed a linker-based blue emission, the (co)doping of Dy^{3+} and Eu^{3+} ions resulted in a single-phase white-light-emitting phosphor [16].

4.2 Titanium MOFs

There are many reports on the synthesis of **titanium-based** MOFs, i.e., in photocatalytic hydrogen evolution, sorption, drug delivery, photoactive materials, and more [17–20]. Martí-Gastaldo and coworkers have described a hydroxamate titanium MOF, which was synthesized from the ligand Benzene-1,4-dihydroxamic acid following a solvothermal technique. The technique indicates that the ligand was suspended in a mixture of 7.2 mL of N,N-dimethylformamide and 2.1 mL of AcOH in a 25 mL Schott bottle. The bottle was sealed and heated in an oven at 120°C for 48 h. MUV-11 is a crystalline, porous material that combines photoactivity with outstanding chemical stability in acid conditions intrinsic to the introduction of siderophore metal binders, **Figure 8** [21]. The structure presents a very distorted octahedral geometry. Sorption of different ions, small and medium-size molecules can result because of

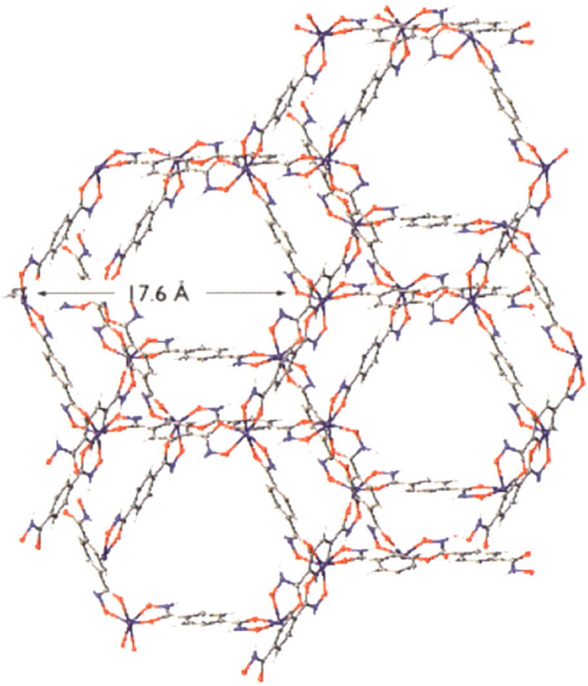

Figure 8.
Pore size representation of a small part of the MOF, dimethyl ammonium molecules were omitted for clarity [21].

the high porosity. The aromatic rings and oxygen atoms present in the structure can interact with different guest molecules via Van der Walls, electrostatic hydrogen bonding, or π-π interactions.

4.3 Vanadium MOFs

While MOFs based on divalent metals, such as Zn^{2+} and Cu^{2+}, have received much attention over the past decade, less progress has been achieved on the synthesis of new MOFs containing tri- and tetravalent metals [22]. Among these, vanadium MOFs are particularly rare. Applications of vanadium-based-MOFs can be found in areas of chemistry such as catalysis, adsorption, separation of N_2, CO_2, CH_4, magnetism, and others [23–25]. Ferey and coworkers reported three-dimensional vanadium (III) dicarboxylate, derived from terephthalic acid in HF, MIL-71. MIL-71 was prepared from a mixture of metallic vanadium, HF, 1,4-benzenedicarboxylic acid, and deionized water heated 3 days at 473 K under hydrothermal conditions. MIL-71 exhibits two features according to the authors: (i), it is the first solid with a two-dimensional inorganic subnetwork among the series of hybrid vanadocarboxylates, and (ii) compared with other trivalent cations, it exemplifies once more the peculiar behavior of V(III), which easily oxidizes into V(IV) [26]. Finally, the Vanadium atoms show an octahedral geometry where four oxygen atoms coordinated to the metal occupy the equatorial positions and two fluorine atoms that occupy the apical positions as observed from **Figure 9**.

More interesting to observe is the pore size obtained in MIL-71 in the different planes. **Figure 9** represents the pores in the MOF, the left side denotes the 010 planes with distances of 3.57 Å × 10.72 Å and the right plane 100 with distances of 3.57 Å × 3.87 Å.

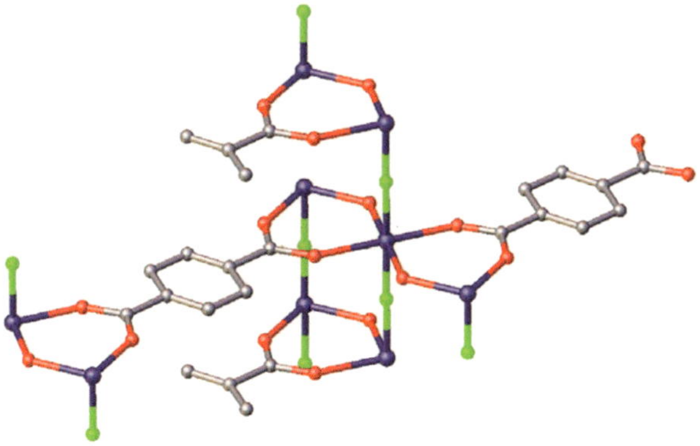

Figure 9.
Molecular representation, where the vanadium atom is hexacoordinated displaying an octahedral geometry.

The right side represents the front part, and the left side represents the lateral view, forming in this way a rectangle. Red color represents oxygen atoms, green color represents fluorine atoms, gray color represents carbon atoms, and purple color represents vanadium atoms. Metals and many small molecules could be adsorbed in this material due to the great number of electrons spported by the oxygen and fluorine atoms and the aromatic rings in the MOF, **Figure 10**.

4.4 Chromium MOFs

Chromium-based MOFs have been widely used with different applications such as environmental remediation, gases sorption, catalysis, carrier systems, among others [27–31]. Feng and coworkers have recently reported an ultrastable High-Connected Chromium Metal-Organic Framework, which was obtained using a Teflon cup, weighing, and combining chromium nitrate nonahydrate, 1,4-benzenedicarboxylic acid, 2,4,6-tri(4-pyridyl)-1,3,5-triazine, and 50 μL hydrofluoric acid dissolved in water. The solvothermal technique indicates that after being stirred for an hour, the vessel was sealed and was subsequently placed in a 220°C preheated oven for 2 days.

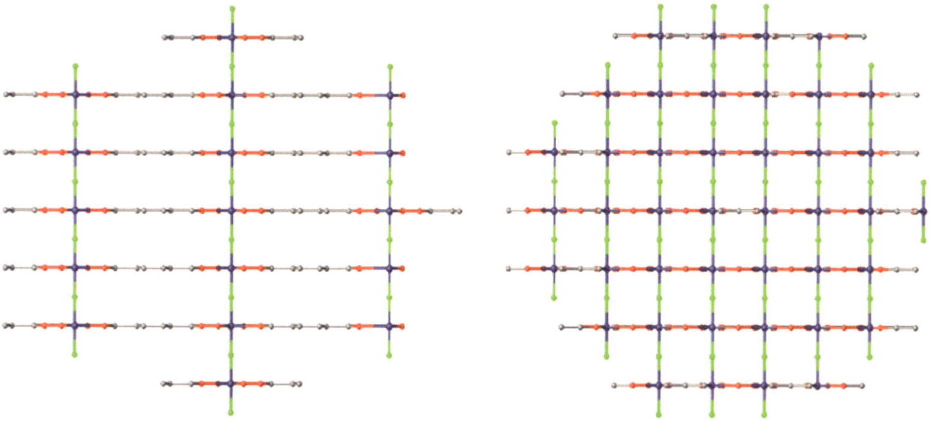

Figure 10.
Lateral view plane 010 (left) and frontal view plane 100 (right) of MIL-71.

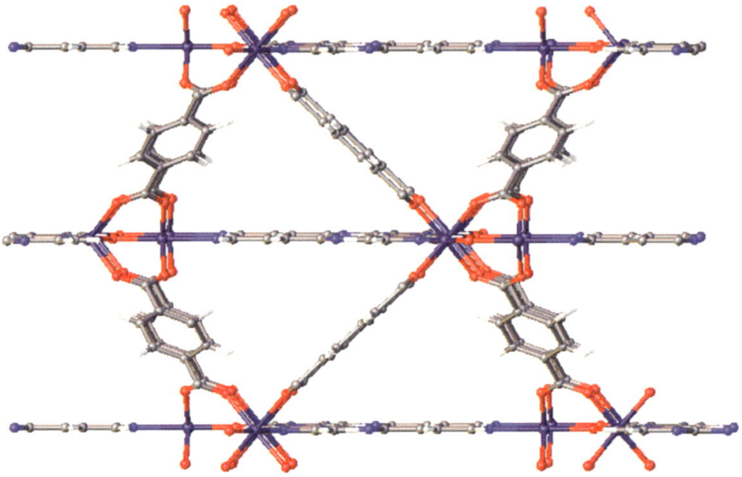

Figure 11.
Illustration showing graphic depicts of the porosity of Cr-based MOF.

As mentioned before, the importance to follow the technique is imperative to obtain the correct product, because a minimal variation of temperature, reaction time, pH, and even stir velocity can conduce to a different product. The obtained MOF gives rise to a tridimensional structure, and the material presents high porosity. The chromium atoms in the structure are hexacoordinated showing an octahedral geometry with typical bond distances in the range from 1.94 Å to 2.17 Å, **Figure 11** [32].

4.5 Manganese MOFs

Manganese-based MOFs are very stables, and the reports in the literature are very large, which include catalysis, materials with magnetic properties, transport, energy, gas sorption, among others [33–37]. Manganese(II) ions/clusters have been used as secondary building units (SUBs) to build MOFs in fields of gas adsorption and magnetism attributed to their specific electronic configuration.

Wang and coworkers reported a manganese-based MOF applied in the decomposition of ozone, which means that the MOF works like a catalytic species to decompose the ozone in water. The technique employed for the synthesis was solvothermal, weighing the corresponding moles of Mn $(NO_3)_2{\cdot}6H_2O$ and the linker H_4TTPE, which were dissolved in a mixture of DMAc/H_2O. The yellow solution obtained was stirred at room temperature for 5 min, sealed in a 25 mL Teflon-lined bomb, and heated at 120°C for 72 h [38].

The material shows according to the X-ray structure, two different kinds of manganese atoms, one manganese being pentacoordinate with a very distorted trigonal bipyramidal geometry, and the second manganese being surrounded by six atoms accommodated in an octahedral fashion. The X-ray shows two different types of pores in the structure, and the water molecules are coordinated to the manganese metal atom. The two water molecules occupy part of the interior of the pores, and although both pores are similar in size, it could be a slight difference in sorption capacity when small molecules go inside the pore and interact either with the benzene electron-rich rings pores or the more exposed electron-rich pores that contain oxygen and nitrogen atoms from the tetrazole moiety, **Figure 12**.

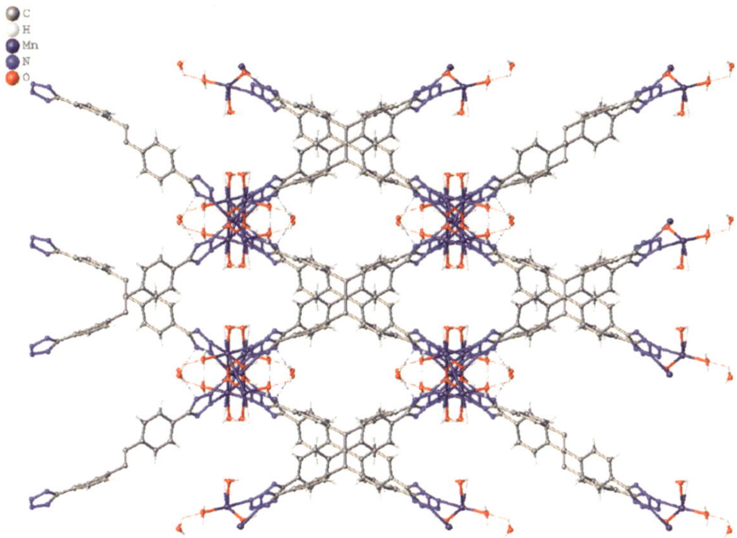

Figure 12.
Manganese-based MOF shows the pores for possible sorption of ions and molecules.

4.6 Iron MOFs

Iron-based MOFs have been widely reported in biological applications due to their high stability, ease of synthesis, and low toxicity of the metal to the human being. Application in areas such as environmental remediation, adsorption of volatile compounds, catalysis, drug delivery, water remedying, glucose biosensing, among others [39–43].

Long and coworkers described an iron MOF that was synthesized from anhydrous ferrous, 1,4-dihydroxyterephthalic acid, DMF, and methanol. The reaction mixture was heated at 393 K and stirred for 18 h to afford a red-orange precipitate. The solid was collected by filtration and washed with 100 mL of DMF to yield 2.0 g (91%) [44].

Although the reaction implies a solvothermal method for 18 h, worth it for the high yielding obtained. In **Figure 12**, we observe a small representation of the MOF, orange, gray, and red spheres represent Fe, C, and O atoms, respectively, and the hydrogen atoms were omitted for clarity. The iron atom possesses two different coordination ways, square pyramidal and octahedral. The pores are well defined and the oxygen atoms possessing two-electron lone pairs can interact with different metals and molecules. Even more, all the channels are well defined, and **Figure 13** shows a small portion of the crystal.

4.7 Cobalt MOFs

Cobalt-based MOFs have been broadly studied due that cobalt salts being cheap and easy to obtain, even more, the cobalt atom can form Penta or Hexa coordinate geometries, increasing the possibilities of coordination modes. These materials found applications in oxygen and hydrogen evolution, magnetism and superconductivity, catalysis, electrocatalysis, synthesis of nanomaterials, and more [45–49].

Burgos and coworkers could develop nanosheets of cobalt MOF for enhanced electrocatalytic water oxidation. The X-ray studies show that there exists just one type of

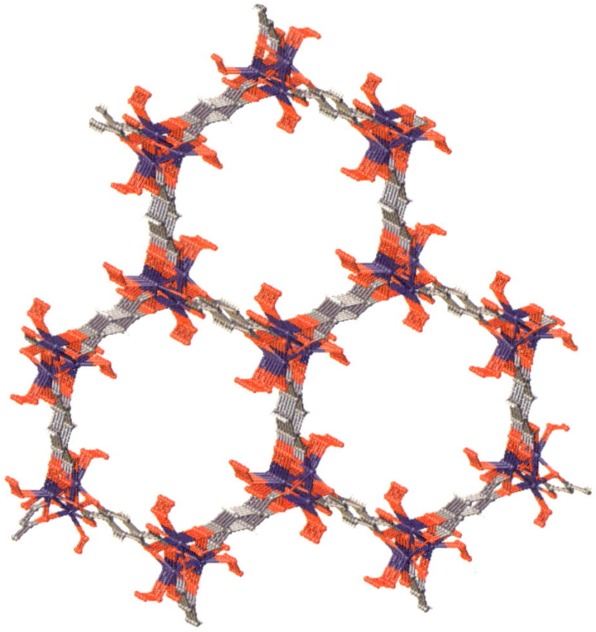

Figure 13.
Well-defined pores in the MOF Fe_2(dobdc), where solvent molecules were omitted for clarity.

cobalt atom, which is hexacoordinate with an octahedral geometry, and the pyridine solvent is coordinated to the cobalt atom. The pyridine rings are close enough to interact via π-π stacking with an average distance of 9.25 Å. The cobalt atoms are hexacoordinated showing a distorted octahedral geometry with coordination distances in the range of 2055–2179 Å. Because this material contains two different types of cavities, both with an electron-rich environment, metals or small molecules can interact in the surroundings to produce stable states [50].

The use of a well-defined cobalt cluster as the starting compound for the synthesis directs the construction of a Co-MOF with an unusual topology. In this MOF, the layered double nanosheets are held together by $\pi - \pi$ stacking interactions between labile pyridine ligands. It has been shown that this material delaminates in the presence of water and that the original 3D layered structure can be regenerated by solvothermal treatment with pyridine so that the individual nanosheets have associated memory **Figure 14.**

4.8 Nickel MOFs

The combination of porosity and the presence of coordinatively unsaturated Ni^{2+} sites are also of special interest because of catalytic properties and the strong H_2 binding affinity. Added to this, the chemical and thermal stability and the presence of accessible Lewis acid sites are some of the reasons why Nickel-based MOFS find application in a great variety of fields of chemistry and reports can be found on catalysis for the ethylene oligomerization, CO_2/CH_4 separation, magnetic and conducting materials, methanol oxidation, among others [51–55].

Gong and coworkers reported the synthesis of Nickel-based MOFS using the microwave-assisted [MA] technique combined with the solvothermal reaction. In contrast to conventional methods, MA methods allow the rapid and systematic investigation of large synthesis fields. This enables the effective discovery of new

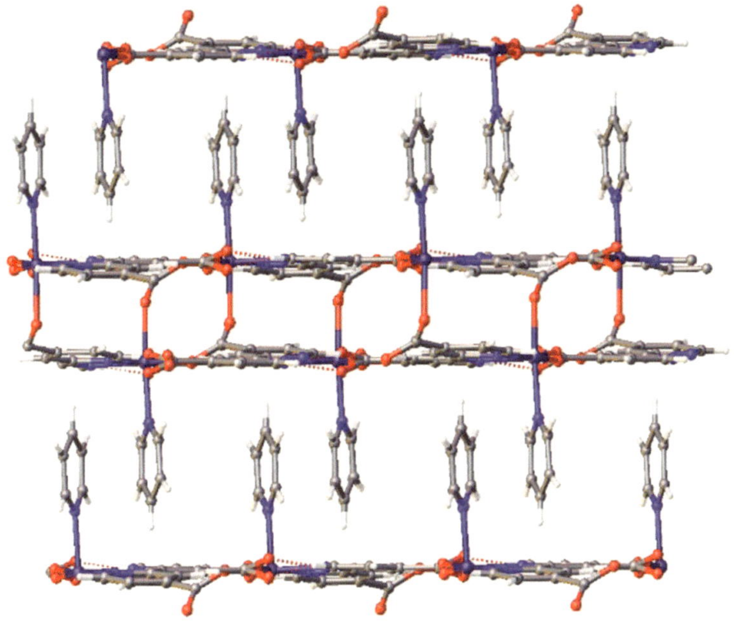

Figure 14.
A small part of the crystal shows the proximity of the pyridine groups that originate the π-π interactions.

compounds, the fast optimization of synthesis conditions, and because of the large amount of data, it allows the extraction of reaction trends. MA methods have been successfully applied in the investigation of porous MOFs.

The resulting material was obtained as a crystalline solid and the X-ray diffraction showed the structure that possesses a hexacoordinated cobalt atom arranged in an octahedral fashion with a Ni-Ni interaction of 2.713 Å. The coordination bond distances are in the range from 1.972 to 2.009 Å [56]. Part of the structure is shown in **Figure 15**.

4.9 Copper MOFs

Reports on MOFs containing copper include novel application in coordination strategies to control the growth orientation of the crystals, modification, and adsorption of different volatile organic compounds, MOF nanoparticles applied for sensitive fluorescent detection of ferric ion, copper-based MOFs for sensitive colorimetric detection of staphylococcus aureus, separation of CO_2 over CH_4 or N_2, among others [57–61].

Park and coworkers adopted a facile green synthesis for the preparation of a copper-based MOF applied in the cycloaddition reaction of CO_2 and epoxide. The coordination geometry around each Cu in the bimetallic cluster is octahedral, where the square base is established by the four BDC units with their four carboxylate oxygen atoms. The PNU-25 structure contains 3D channels with rectangular windows of dimensions 15.21 × 10.80, which undergo interpenetration *via* various supramolecular interactions forming an overall triple interpenetrated network, as depicted in **Figure 16** [62].

The heterogeneous PNU catalysts efficiently catalyzed the synthesis of cyclic carbonates by the coupling of epoxide and CO_2 under ambient pressure and lower reaction temperature. The PNU catalysts demonstrated remarkably good thermal stability for the cycloaddition reaction. The coordinatively unsaturated Cu(II) units

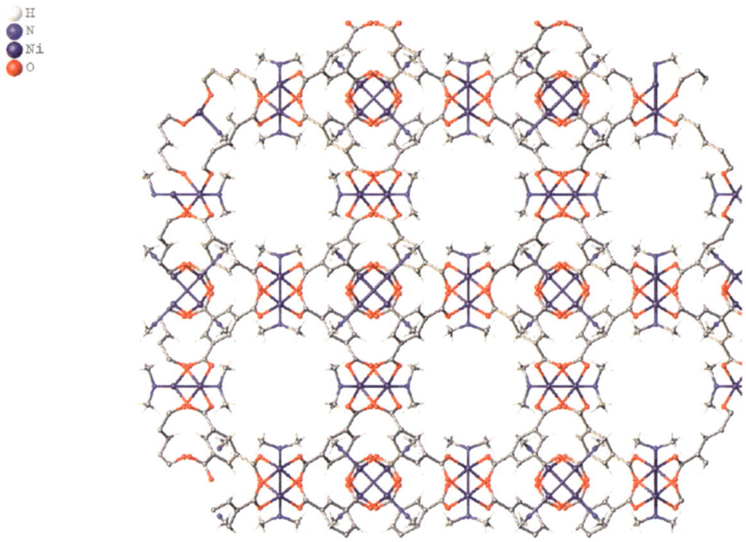

Figure 15.
Part of the structure of Ni-based MOF obtained by microwave-assisted technique.

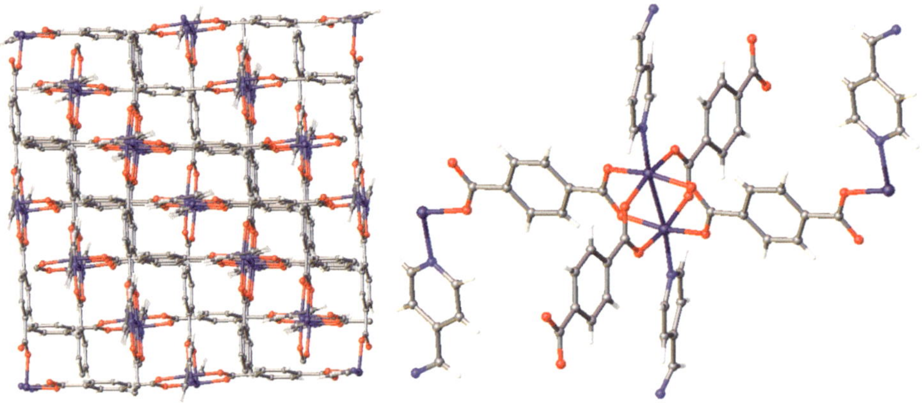

Figure 16.
The PNU-25 structure contains 3D channels left, with copper metal atoms being hexacoordinated within an octahedral geometry, right.

and the basic N atoms resulted in a large number of acidic-basic sites, facilitating the conversion of epoxides.

4.10 Zinc MOFs

One of the important applications of zinc-based MOFs is the separation of materials, gases, and compounds. By using the modified MOFs, diverse gases, organic and inorganic compounds were separated, i.e., H_2/CO_2, Xe, and Kr, $O_2/N_2/CH_4$, other applications such as catalysis in organic synthesis, removal and detection of antibiotics in water, sensing, photocatalytic activity, semiconductive and magnetic properties, thin-film nanocomposite membrane incorporated with Porous Zn-Based Metal-Organic Frameworks, among others [63–67].

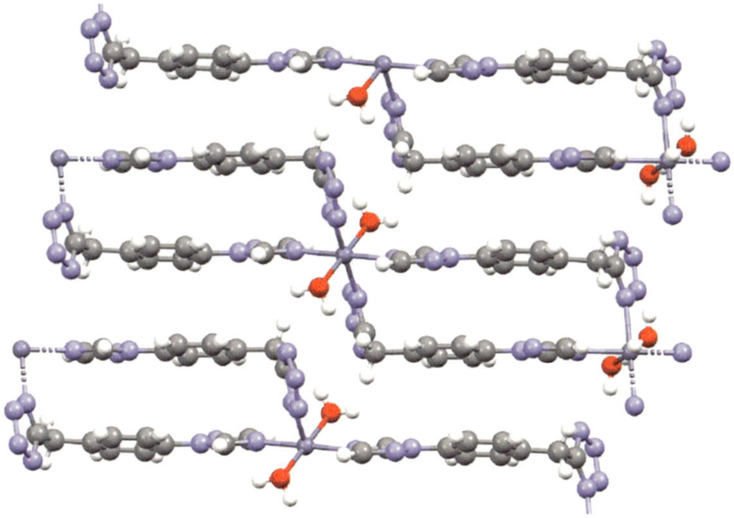

Figure 17.
A small fraction of Zn-ttb-bdc MOF shows two coordination water molecules.

Yang and coworkers develop a low toxicity MOF for the detection of organic and inorganic contaminants from water. The material was obtained according to the mixture of the linker Httb and Zn $(NO_3)_2 \cdot 6H_2O$ in a mixture of solvents, H_2O, and CH_3CN, and the mixture was stirred for 20 min before adding in a steel vessel at 160°C for 5 days to obtain the Zn-based MOF, with yield 64%.

The MOF exhibits high water and chemical stability as well as excellent fluorescence properties. The remaining binding sites show higher sensitivity and better fluorescence response to the representative organic micropollutant TNP and inorganic pollutants (Fe^{3+} and $Cr_2O_7^{2-}$ in wastewater. The Zn metal is hexacoordinated to four nitrogen atoms from the tetrazole ring and two oxygen atoms belonging to water molecules, **Figure 17** [68]. The final structure presents π-π interactions between the aromatic rings.

5. Synthesis OF MOFs

There are many reports in the literature about the ways and techniques commonly used, frequently called reaction conditions, which are in some cases very specific to synthesize the metal transition-based MOFs. The different synthesis methods that have been applied in the last 20 years in addition to room temperature synthesis are conventional electric (CE) heating, microwave (MW) heating, electrochemistry (EC), mechanochemistry (MC), and ultrasonic (US) methods that have been employed.

Conventional step-by-step methods, as well as high-throughput methods, have been employed in some of the studies. In addition to the crystal size or shape, thin films, membranes, and various other shapes made of MOFs have been reported, which require the application of different and specific synthesis methods. Nevertheless, many other synthetic methods and parameters, such as temperature, reaction time, pressure, pH, and solvent, must be considered as well. Numerous different synthetic approaches, including slow diffusion, hydrothermal, and solvothermal, can be applied to produce MOFs relying on the resulting structures and features [69].

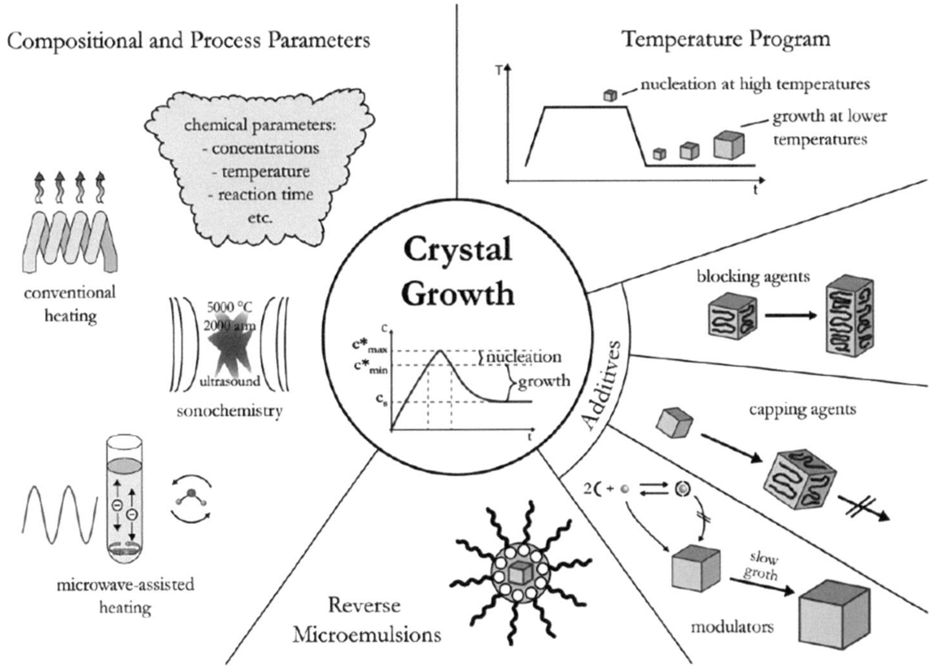

Figure 18.
Strategies and methods for the synthesis of crystalline materials [2].

Figure 18 presents a resume of the different techniques applied for the synthesis of MOFs, where we observe the conventional heating, sonochemistry, microwave-assisted, using different temperature programs that let the correct nucleation and growth of single crystals.

6. Environmental applications of transition metal-based MOFs

As could be discussed in the previous part, there are many different techniques to achieve the synthesis of MOFs, and something very important to consider is the combination of procedures that can allow the formation of materials with specific pore sizes and shapes. The environmental applications of MOFs depend on the final structure obtained, which is very related to the technique employed and the metals and linkers used for the synthesis. Pore size and electron charge distribution into the pores and the surface let those metals or even neutral or charged molecules interact with each other. In this section, we'll focus on three environmental applications, heavy metal atoms, and azo dyes removal from water, hydrogen evolution, and catalysis directed to the transesterification process of non-edible oils to obtain biofuels.

6.1 Metals and azo dyes removal from wastewater

There are many reports in the literature of heavy metal atoms removed from aqueous mediums using crystalline materials. The remotion of those metals from wasted water is due to the weak interactions generated between the metals and the electron-rich environment of the pores. In the same way, the remotion of azo dyes (medium-size molecules). i.e., Congo red, methylene blue, and methyl orange from wastewater, has been widely described, due to those azo dyes are considered toxic and carcinogenic pullulans in the water.

Heavy metals such as arsenic, cadmium, chromium, copper, lead, mercury, nickel, and zing with a density over 5 g cm^{-3}polluting our water are a rapidly growing global concern. These elements can be found within the environment, be it in water reservoirs, the atmosphere, or soil [70]. The recovery and elimination of toxic metal ions from wastewater are of concern with increasing awareness toward the need for protecting nature. Mercury and lead are the most toxic species and can cause bioaccumulation in kidneys, brain, lung tissues, gastrointestinal tract, central nervous system, and reproductive system [71].

To date, several approaches have been applied to remove the heavy metals from water including nanofiltration, membrane separation, ion exchange, resin, photocatalytic degradation, chemical moisture, membrane filtration, freezing, chemical deposition, biological treatment, reverse osmosis, adsorption, etc. [72].

Studies to investigate the sorption mechanisms governing the sorption process and to determine whether the sorption mechanism are controlled by a chemical or physical mechanism, Morsali et al. applied conventional sorption kinetics models such as pseudo-first-order, pseudo-second-order, and intraparticle diffusion to the study. The correlation coefficient of the pseudo-second-order model for Cd(II) is close to 1 (R^2 = 0.9999) and exhibited that the pseudo-second-order model is more consistent with experimental data than the other model. Therefore, it is compatible with chemical sorption, which indicates that transfer, exchange, or sharing of electrons has taken place. Therefore, supramolecular interactions between cadmium ion and the free electrons of ligand nitrogens and the electrostatic interaction between the cation and the dipole of nitrogen on the dihydropyrazine ring of ligand are proposed as the dominant interactions between the metal ion and the structure [73].

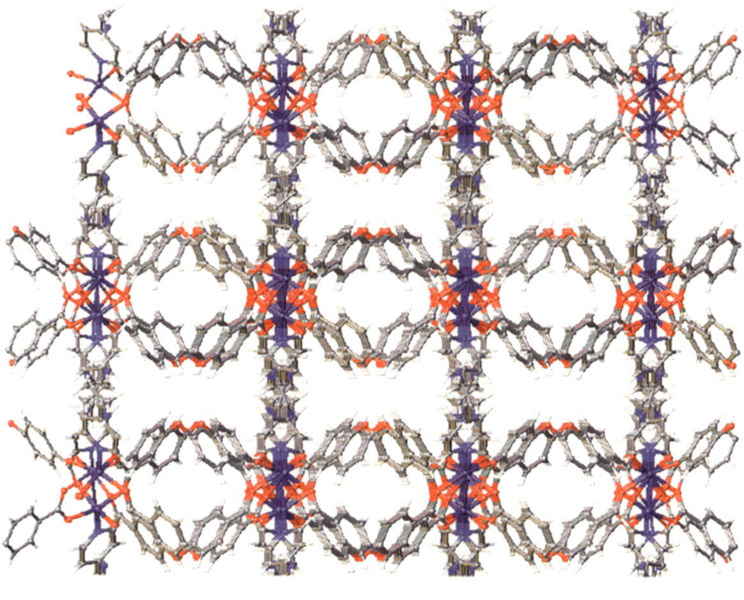

Figure 19.
A small portion in MOF TMU-60 and pore size obtained after the synthesis.

The X-ray structure showed that the aromatic rings from the linker are available to interact through supramolecular interactions with different metals or even small molecules through week interactions **Figure 19**.

The synthesis was achieved in DMF at 60°C, and then the solution was transferred to a Teflon autoclave and heated to 120°C for 72.

Large amounts of dyes are commonly used in many manufacturing, such as textiles, clothing, and printing. Among these, azo dyes are one of the largest groups that are heavily used, but most difficult to be degraded, and even more, the azo dyes and their derivative products are toxic to the aquatic environment and are mutagenic and carcinogenic to humans. Therefore, the treatment of wastewaters containing these dyes is necessary [74].

Numerous studies have considered new processes to eliminate emerging organic contaminants (EOCs) from water, i.e., ozonization, chlorination, sonodegradation, biodegradation, inorganic heterogeneous catalysis, activated carbon treatment, and more. Among these technologies, metal-organic frameworks (MOFs) have been recently investigated for the removal of contaminants in water [75].

Separation or degradation requires strong research efforts to modify MOFs via controlling their pore diameter (adding functional groups, creating defects) or to construct MOF composites to engineer materials as improved adsorbents and/or catalysts for contaminated water treatment. MOFs and MOF composites have been proposed for the removal of a wide range of contaminants, including dyes, pharmaceuticals, plasticizers, herbicides and pesticides, industrial products, among others. While a few reviews have already documented the use of MOFs in the removal of novel contaminants [76].

Choe et al. reported the adsorptive removal of various dyes using a Zr porphyrinic MOF, PCN-224. The plausible mechanism for adsorptive removal revealed multiple interactions between the dye and porphyrin linker/Zr6 node via $\pi - \pi$ interactions and hydrogen bonding, respectively. Such results demonstrate that PCN-224 is an excellent adsorbent, providing superior water stability, pore aperture of suitable size, and multiple interaction sites **Figure 20** [77].

The best performance of dye adsorption onto PCN-224 comes from structural properties such as appropriate pore aperture, volume as well as various types of

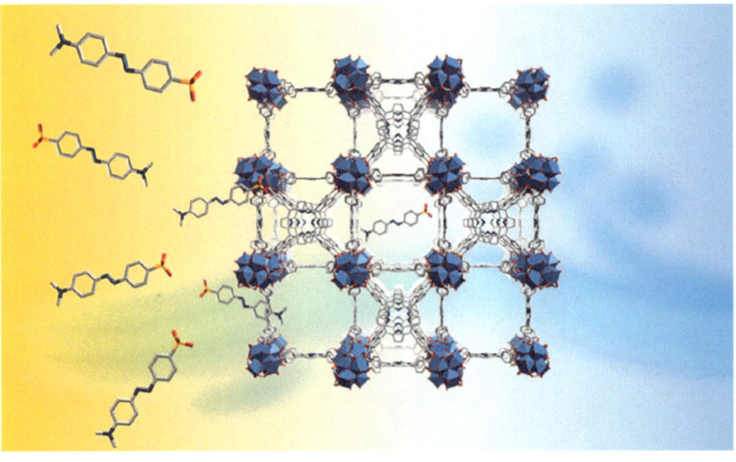

Figure 20.
Zr porphyrinic metal-organic framework (MOF) for removal of azo dyes [77].

interactions such as π–π interaction, hydrogen bonding, and electrostatic interaction. Significantly, an interaction between the sulfonate group of the MO molecule and the Zr6 node of PCN-224 was demonstrated through experimental and theoretical studies. Arrangement of tripod form between Zr node and sulfonate group of the MO molecule can form hydrogen bonding.

6.2 Hydrogen evolution

Highly efficient hydrogen evolution reactions (HERs) will determine the mass distributions of hydrogen-powered clean technologies in the future. That's why hydrogen evolution is one of the topics more explored in the last years because molecular hydrogen is the best environmentally friendly fuel available, which reacts with oxygen to produce energy and water.

The procedures employed to produce hydrogen are still expensive because of the catalysts employed to obtain the gas and the conditions and hence the high-cost obtention. Current industrial hydrogen production methods include coal gasification (followed by water-gas shift reaction), steam reforming, cryogenic distillation, and water splitting.

The specific surface area of MOFs can range from 1000 to over 6000 $m^2 g^{-1}$, thanks to their tailorable porous structures, which play a significant role in enhancing the catalytic HER process. Large surface area and pore volume ensure sufficient contact between the electrolyte (or reactant solution) and the surface of the catalyst, which essentially improves catalytic performance by exposing more active sites for the catalytic reactions to take place. It is that research on MOFs for HER primarily focuses on the following three techniques: electrocatalytic, photocatalytic, and chemocatalytic HER [78].

MOFs, MOF supports, and MOF derivatives can be utilized as catalysts in the abovementioned hydrogen production methods. Most of the materials used for photocatalytic hydrogen evolution PHE applications include inorganic oxides such as TiO_2, ZnO, and $SrZrO_3$, due to their high stability. The application of metalorganic frameworks in these processes is limited due to the loss of efficiency attributed to the recombination of electrons and holes [79]. Even so, together with the unique porous structure of MOFs, a remarkable hydrogen evolution reaction HER performance can be achieved using different overpotential in phosphate buffer solution (PBS, pH = 7.0) [80].

In the same way, the significant effect of crystallinity in the photocatalytic activity of metal-organic frameworks was demonstrated through the evaluation of different samples with different crystallinity in the HER reactions. The samples with high crystallinity produce too a higher amount of hydrogen, which is attributed to the lower recombination supported by the experiments of photoluminescence and electrochemical impedance directly related to a high-ordered material.

Rivera et al. described the synthesis of a BDC-Zn MOF, which was firstly used for methyl orange and methyl blue sorption and followed by PHE under solar light. MO presented the best adsorption result, with a maximum adsorption capacity of 2100 mg/g, which is higher than all the MOFs reported in the literature. For HER, the activity was enhanced 24 times in photocatalyst with MO adsorbed, and 27 times for the MB adsorbed (from 47 to 1148 and 1259 μmol/gh, respectively). This result is attributed to better light adsorption and a decrease in charge recombination. It's important to mention that even though the reflux method presented the disadvantage that it is not possible to obtain single crystals, the reaction conditions such as

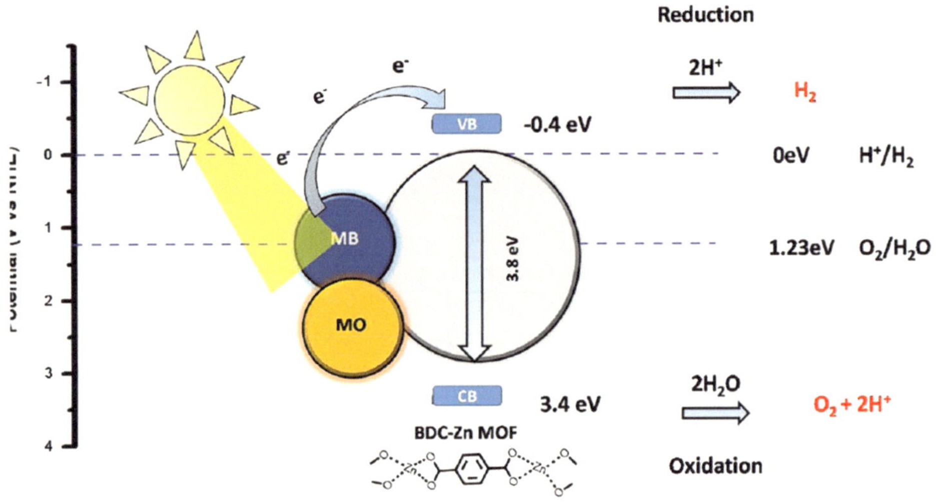

Figure 21.
Mechanism of HER by using the BDC-Zn-MOF obtained under solvothermal conditions [81].

temperature, pressure, and time are more ecological in comparison with the traditional MOF 5 solvothermal synthesis and analogous [81].

Figure 21 shows the mechanism for HER. In the first step, MO/MB molecules linker to BDC-Zn MOF captures the light from the solar simulator. Light produces the transference of electrons from the conduction (LUMO orbital analog) to the valence band (HOMO orbital analog), this produces a hole formed in the conduction band, which oxidizes water molecules, finally, the electron in the valence band reduces H^+ to H_2. At the same time the species O_2^- and OH^- are formed, and these could affect MOFs by redox reactions; however, possibly MO/MB acts as a sacrificial agent avoiding the BDC-Zn MOF structure collapse by the capture of these species. Therefore, growing hydrogen production is observed.

6.3 Catalytic transesterification reaction

Biodiesel is green and renewable energy, which is a promising substitute to replace fossil combustibles. Normally, biodiesel is produced via transesterification/esterification with the assistance of a homogeneous or heterogeneous catalyst. Biodiesel is a series of fatty acid alkyl ester (FAAE), which is normally derived from transesterification of vegetable oil or animal fat (triglyceride) and alcohol with the assistance of catalyst as shown in **Figure 22**.

In general, transesterification is the most used technique for biodiesel production, which is proceeded in three steps. In the first step, triglyceride reacts with alcohol generating monomolecular FAAE and diglyceride. Then, diglyceride reacts with alcohol resulting in monomolecular FAAE and monoglyceride. Finally, monoglyceride reacts with alcohols giving rise to monomolecular FAAE and glycerol [82].

Industrial biodiesel production is based on the transesterification of triglycerides with methanol or ethanol using stoichiometric amounts of strong Brönsted base (e.g., sodium or potassium hydroxide/methoxide) and homogeneous acid catalysis, such as H_2SO_4, HCl, H_3PO_4, but the problem with this homogeneous catalyst is their difficult recovery from the media. The ease of heterogeneous catalyst to be recovered at the

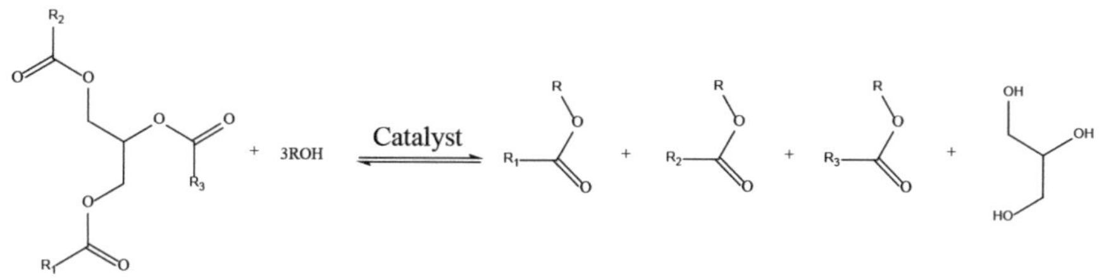

Figure 22.
Transesterification reaction of triglycerides in alcohol as a solvent.

end of the reaction just by a simple filtration reduces dramatically the cost in the biodiesel obtention process.

One of the properties of MOFs compared with other porous solids is the facile introduction of the desired active sites. Solvent/water molecules weakly bounded around metal nodes in MOFs could be removed upon activation; thus, these coordinatively unsaturated metal sites behave as Lewis acid sites [83].

MOFs can be used in combination with other materials to form composites or as catalytic support to solve two specific problems, stability, and recyclability issues. The porous materials, metal-organic frameworks (MOFs) are suitable for enzyme immobilization owing to their ideal features of tunable pore size, and topological and compositional versatility. Currently, multiple enzyme-MOF composites have been successfully created using physical adsorption or covalent attachment strategies. Also, because MOFS catalyze many reactions, they accelerate the velocity of the reaction and low the transition states of the molecules to react faster and slower reaction times and energy consumption [84].

To increase the catalytic activity in MOFs, three different strategies can be used. The first strategy is post modification of MOFs, which results in the linker modification to increase the active sites and hence the catalytic activity. The second strategy is the formation of composites, which results in mixing the MOFs with different basic or acid materials together to obtain better results and structural stability in MOFs. The third strategy is to develop new linkers that incorporate the specific active sites that remain free once the MOFs are obtained [85]. Finally, because the transesterification process requires very large reactions times, the energy consumption is high; therefore, different synthesis techniques can be used in combination. Rivera et al. could obtain through hydrothermal synthesis a 3D-Co-based MOF. The Ultrasonic-assisted synthesis is a powerful technique, the yields obtained are increased, and the reaction times are decreased using milder conditions.

Rivera et al. reported the green synthetic approach that involves the ultrasonic technique that allows in terms of energy conservation higher purity products. The high catalytic activity exhibited by the catalyst reported has to do in part with the free carboxylic acid groups that remain free in the structure, as revealed by the X-ray diffraction. The application of this catalyst resulted in 80% of total conversion after 12 h at just 60°C [86].

The use of composite MOFs is widely described in the literature, and the idea to produce biodiesel is to add new basicity or acidity sites to the material to increase the catalytic activity or simply to get better stability. Zhang et al. synthesized the reusable and highly active Fe-BTC and UiO-66 metal-organic framework by hydrothermal method, and the MOF was applied for the acid-catalyzed esterification of oleic acid with methanol. Typically, the esterification reactions of oleic acid with methanol

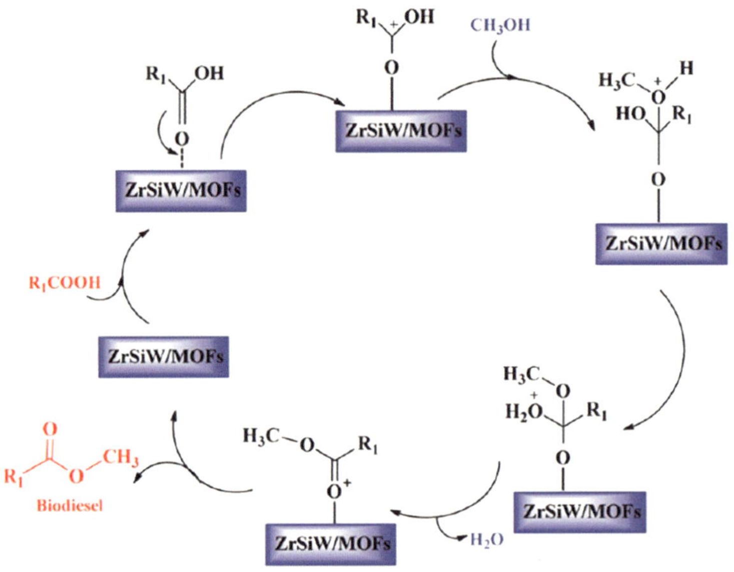

Figure 23.
Possible reaction mechanism for the esterification reaction process [87].

were performed in a 50 mL stainless steel reactor equipped with a magnetic stirrer. The mechanism proposed for the esterification reaction is shown in **Figure 23**.

Characterization analyses indicated that the ZrSiW/UiO-66 possesses an appropriate structure and high acidity. The highest oleic acid conversion of 98.0% was obtained using the ZrSiW/UiO-66 nano-hybrids nanocatalyst under the optimal esterification reactions: 150°C, 0.24 g catalyst, 1:20 molar ratio of oleic acid to methanol, and a reaction time of 4 h [87]. The easy preparation of MOF composites lets these materials find better opportunities to be investigated [87].

7. Conclusions

Metal-organic materials are easy to synthesize and have a wide variety of applications. MOF materials are constructed from linkers and different metal atoms, which can interact via coordination bonds to produce 1D, 2D, and 3D porous structures. The final porous structure depends primarily on the linkers and metals used for the synthesis but equally important to obtain specific crystalline structures and hence high-ordered and porous materials is to apply the correct technique that involves the correct pressure, reaction time, temperature, pH value, solvent, or a mixture of solvents, stirring, or even microwave and ultrasonic-assisted synthesis, which can be applied and combined with other techniques to obtain desirable size and porous structures in the MOFs. The environmental applications of MOFs such as sorption of metals and azo dyes depend on the final structure obtained, the pore size, and the electron charge distribution into the pores, and the surface allows metals or even neutral or charged molecules to interact with the MOF to form different contacts such as hydrogen bonding, π-π interactions, or simply interact through Van der walls forces. On the other hand, the catalytic activity for the transesterification process to obtain biodiesel depends mainly on the availability of metal centers, which are acid sites that can be increased by adding more acid sites by a post-modification synthesis or simply by using a specific linker containing the acid sites in the original structure. Finally, the hydrogen evolution reaction depends mainly on the structure of MOFs (crystallinity) and the linkers, that is, delocalized systems that allow HOMO and LUMO orbitals to get less separated, and hence the electrons can get transferred more efficiently to oxidase water.

Acknowledgements

Thanks to committee pro-mejoras from Universidad Veracruzana, Facultad de Ciencias Químicas, Orizaba, Ver. México, who supported the funding of this project.

Conflict of interest

"The authors declare no conflict of interest."

Author details

Lidia E. Chiñas-Rojas[1,2], Guadalupe Vivar-Vera[1], Yafeth F. Cruz-Martínez[1], Seth Limón Colohua[1], José María Rivera[1] and Eric Houbron[1*]

1 Facultad de Ciencias Químicas, Universidad Veracruzana, Orizaba, Veracruz, Mexico

2 Universidad Martí, Poza Rica, Veracruz, Mexico

*Address all correspondence to: ehoubron@uv.mx

References

[1] Seth S, Matzger AJ. Metal–organic frameworks: Examples, counterexamples, and an actionable definition. Crystal Growth & Design. 2017;**17**:4043

[2] Stock N, Biswas S. Synthesis of metal-organic frameworks (MOFs): Routes to various MOF topologies, morphologies, and composites. Chemical Reviews. 2012;**112**:933

[3] Desai AV, Sharma S, Let S, Ghosh SK. N-donor linker based metal-organic frameworks (MOFs): Advancement and prospects as functional materials. Coordination Chemistry Reviews. 2019;**395**:146-192

[4] Wang CC, Yang CH, Tseng SM, Lee GH, Chiang YP, Sheu HS. Self-assembly of a mixed-ligand metal-coordination polymeric network of cadmium (II) croconate with 4, 4′-Bipyridine. Inorganic Chemistry. 2003;**42**(25):8294-8299

[5] Arenzano JA, del Campo JM, Virues JO, Ramirez-Montes PI, Santillán R, Rivera JM. Theoretical study of the hydrogen bonding interaction between Levodopa and a new functionalized pillared coordination polymer designed as a carrier system. Journal of Molecular Structure. 2015;**1083**:106-110

[6] Goura J, Chandrasekhar V. Molecular metal phosphonates. Chemical Reviews. 2015;**115**(14):6854-6965

[7] Ghasempour H, Wang KY, Powell JA, ZareKarizi F, Lv XL, Morsali A, et al. Metal–organic frameworks based on multicarboxylate linkers. Coordination Chemistry Reviews. 2021;**426**:213542

[8] Lu W, Wei Z, Gu ZY, Liu TF, Park J, Park J, et al. Tuning the structure and function of metal–organic frameworks via linker design. Chemical Society Reviews. 2014;**43**(16):5561-5593

[9] Ding M, Cai X, Jiang HL. Improving MOF stability: Approaches and applications. Chemical Science. 2019;**10**(44):10209-10230

[10] Yang D, Chen Y, Su Z, Zhang X, Zhang W, Srinivas K. Organic carboxylate-based MOFs and derivatives for electrocatalytic water oxidation. Coordination Chemistry Reviews. 2021;**428**:213619

[11] Alnaqbi MA, Alzamly A, Ahmed SH, Bakiro M, Kegere J, Nguyen HL. Chemistry and applications of s-block metal–organic frameworks. Journal of Materials Chemistry A. 2021;**9**(7):3828-3854

[12] Pillai RS, Benoit V, Orsi A, Llewellyn PL, Wright PA, Maurin G. Highly selective CO_2 capture by small pore scandium-based metal–organic frameworks. The Journal of Physical Chemistry C. 2015;**119**(41):23592-23598

[13] Dixit M, Adit Maark T, Ghatak K, Ahuja R, Pal S. Scandium-decorated MOF-5 as potential candidates for room-temperature hydrogen storage: A solution for the clustering problem in MOFs. The Journal of Physical Chemistry C. 2012;**116**(33):17336-17342

[14] Rönfeldt P, Reinsch H, Poschmann MPM, Terraschke H, Stock N. Scandium metal–organic frameworks containing tetracarboxylate linker molecules: Synthesis, structural relationships, and properties. Crystal Growth & Design. 2020;**20**(7):4686-4694

[15] Rönfeldt P, Reinsch H, Faßheber N, Terraschke H, Stock N.

Synthesis and characterization of a layered scandium MOF containing a sulfone-functionalized V-shaped linker molecule. European Journal of Inorganic Chemistry. 2020;**2020**(13):1147-1152

[16] Rönfeldt P, Grape ES, Inge AK, Novikov DV, Khadiev A, Etter M, et al. A scandium MOF with an unprecedented inorganic building unit, delimiting the micropore windows. Inorganic Chemistry. 2020;**59**(13):8995-9004

[17] Wang J, Cherevan AS, Hannecart C, Naghdi S, Nandan SP, Gupta T, et al. Ti-based MOFs: New insights on the impact of ligand composition and hole scavengers on stability, charge separation and photocatalytic hydrogen evolution. Applied Catalysis B: Environmental. 2021;**283**:119626

[18] Chen X, Peng X, Jiang L, Yuan X, Yu H, Wang H, et al. Recent advances in titanium metal–organic frameworks and their derived materials: Features, fabrication, and photocatalytic applications. Chemical Engineering Journal. 2020;**395**:125080

[19] Nguyen HL. The chemistry of titanium-based metal–organic frameworks. New Journal of Chemistry. 2017;**41**(23):14030-14043

[20] Xie Y, Liu X, Ma X, Duan Y, Yao Y, Cai Q. Small titanium-based MOFs prepared with the introduction of tetraethyl orthosilicate and their potential for use in drug delivery. ACS Applied Materials & Interfaces. 2018;**10**(16):13325-13332

[21] Padial NM, Castells-Gil J, Almora-Barrios N, Romero-Angel M, Da Silva I, Barawi M, et al. Hydroxamate titanium–organic frameworks and the effect of siderophore-type linkers over their photocatalytic activity. Journal of the American Chemical Society. 2019;**141**(33):13124-13133

[22] Zhuang H, Yang F. Vanadium metal-organic framework derived 2D hierarchical VO2 nanosheets grown on carbon cloth for advanced flexible energy storage devices. Surfaces and Interfaces. 2021;**25**:101232

[23] Van Der Voort P, Leus K, Liu YY, Vandichel M, Van Speybroeck V, Waroquier M, et al. Vanadium metal–organic frameworks: Structures and applications. New Journal of Chemistry. 2014;**38**(5):1853-1867

[24] Phan A, Czaja AU, Gándara F, Knobler CB, Yaghi OM. Metal–organic frameworks of vanadium as catalysts for conversion of methane to acetic acid. Inorganic Chemistry. 2011;**50**(16):7388-7390

[25] Yan Y, Luo Y, Ma J, Li B, Xue H, Pang H. Facile synthesis of vanadium metal-organic frameworks for high-performance supercapacitors. Small. 2018;**14**(33):1801815

[26] Barthelet K, Adil K, Millange F, Serre C, Riou D, Férey G. Synthesis, structure determination and magnetic behaviour of the first porous hybrid oxyfluorinated vanado (iii) carboxylate: MIL-71 or $V^{III}_2(OH)_2F_2\{O_2C\text{-}C_6H_4\text{-}CO_2\}$ H_2O, K. Journal of Materials Chemistry. 2003;**13**(9):2208-2212

[27] Park J, Feng D, Zhou HC. Dual exchange in PCN-333: A facile strategy to chemically robust mesoporous chromium metal–organic framework with functional groups. Journal of the American Chemical Society. 2015;**137**(36):11801-11809

[28] Gargiulo N, Peluso A, Aprea P, Hua Y, Filipović D, Caputo D, et al. A chromium-based metal organic framework as a potential high performance adsorbent

for anaesthetic vapours. RSC Advances. 2014;**4**(90):49478-49484

[29] Ren J, Dyosiba X, Musyoka NM, Langmi HW, North BC, Mathe M, et al. Green synthesis of chromium-based metal-organic framework (Cr-MOF) from waste polyethylene terephthalate (PET) bottles for hydrogen storage applications. International Journal of Hydrogen Energy. 2016;**41**(40):18141-18146

[30] Bhattacharjee S, Chen C, Ahn WS. Chromium terephthalate metal–organic framework MIL-101: Synthesis, functionalization, and applications for adsorption and catalysis. RSC Advances. 2014;**4**(94):52500-52525

[31] Zhong G, Liu D, Zhang J. Applications of porous metal–organic framework MIL-100 (M)(M= Cr, Fe, Sc, Al, V). Crystal Growth & Design. 2018;**18**(12):7730-7744

[32] Yang H, Peng F, Hong AN, Wang Y, Bu X, Feng P. Ultrastable high-connected chromium metal–organic frameworks. Journal of the American Chemical Society. 2021;**143**(36):14470-14474

[33] López-Cabrelles J, Mañas-Valero S, Vitórica-Yrezábal IJ, Šiškins M, Lee M, Steeneken PG, et al. Chemical design and magnetic ordering in thin layers of 2D metal–organic frameworks (MOFs). Journal of the American Chemical Society. 2021;**143**(44):18502-18510

[34] Liu L, Li L, DeGayner JA, Winegar PH, Fang Y, Harris TD. Harnessing structural dynamics in a 2D manganese–benzoquinoid framework to dramatically accelerate metal transport in diffusion-limited metal exchange reactions. Journal of the American Chemical Society. 2018;**140**(36):11444-11453

[35] Shinde PA, Seo Y, Lee S, Kim H, Pham QN, Won Y, et al. Layered manganese metal-organic framework with high specific and areal capacitance for hybrid supercapacitors. Chemical Engineering Journal. 2020;**387**:122982

[36] Ladrak T, Smulders S, Roubeau O, Teat SJ, Gamez P, Reedijk J. Manganese-based metal–organic frameworks as heterogeneous catalysts for the cyanosilylation of acetaldehyde. European Journal of Inorganic Chemistry. 2010;**24**:3804

[37] Singh K, Guillen Campos JDJ, Dinic F, Hao Z, Yuan T, Voznyy O. Manganese MOF enables efficient oxygen evolution in acid. ACS Materials Letters. 2020;**2**(7):798-800

[38] Sun ZB, Si YN, Zhao SN, Wang QY, Zang SQ. Ozone decomposition by a manganese-organic framework over the entire humidity range. Journal of the American Chemical Society. 2021;**143**(13):5150-5157

[39] Liu X, Zhou Y, Zhang J, Tang L, Luo L, Zeng G. Iron containing metal–organic frameworks: Structure, synthesis, and applications in environmental remediation. ACS Applied Materials & Interfaces. 2017;**9**(24):20255-20275

[40] Wang K, Feng D, Liu TF, Su J, Yuan S, Chen YP, et al. A series of highly stable mesoporous metalloporphyrin Fe-MOFs. Journal of the American Chemical Society. 2014;**136**(40):13983-13986

[41] Liu X, Liang T, Zhang R, Ding Q, Wu S, Li C, et al. Iron-based metal–organic frameworks in drug delivery and biomedicine. ACS Applied Materials & Interfaces. 2021;**13**(8):9643-9655

[42] Joseph J, Iftekhar S, Srivastava V, Fallah Z, Zare EN, Sillanpää M. Iron-based metal-organic framework:

Synthesis, structure and current technologies for water reclamation with deep insight into framework integrity. Chemosphere. 2021;**284**:131171

[43] Fan D, Chen C, Lu S, Li X, Jiang M, Hu X. Highly stable two-dimensional iron monocarbide with planar hypercoordinate moiety and superior Li-ion storage performance. ACS Applied Materials & Interfaces. 2020;**12**(27):30297-30303

[44] Wang D, Li Z. Iron-based metal–organic frameworks (MOFs) for visible-light-induced photocatalysis. Research on Chemical Intermediates. 2017;**43**(9):5169-5186

[45] Tripathy RK, Samantara AK, Behera JN. A cobalt metal–organic framework (Co-MOF): A bi-functional electro active material for the oxygen evolution and reduction reaction. Dalton Transactions. 2019;**48**(28):10557-10564

[46] Dong JL, Xie F, Du JQ, Lan HM, Yang RX, Wang DZ. Cobalt MOFs base on benzimidazol and varied carboxylate ligands with higher capacitance for supercapacitors and magnetic properties. Journal of Solid State Chemistry. 2019;**279**:120917

[47] Zhou P, Wan J, Wang X, Xu K, Gong Y, Chen L. Nickel and cobalt metal-organic-frameworks-derived hollow microspheres porous carbon assembled from nanorods and nanospheres for outstanding supercapacitors. Journal of Colloid and Interface Science. 2020;**575**:96-107

[48] Cai X, Peng F, Luo X, Ye X, Zhou J, Lang X, et al. Understanding the evolution of cobalt-based metal-organic frameworks in electrocatalysis for the oxygen evolution reaction. ChemSusChem. 2021;**14**(15):3163-3173

[49] Gutiérrez-Tarriño S, Olloqui-Sariego JL, Calvente JJ, Palomino M, Minguez Espallargas G, Jorda JL, et al. Cobalt metal–organic framework based on two dinuclear secondary building units for electrocatalytic oxygen evolution. ACS Applied Materials & Interfaces. 2019;**11**(50):46658-46665

[50] Gutiérrez-Tarriño S, Olloqui-Sariego JL, Calvente JJ, Espallargas GM, Rey F, Corma A, et al. Cobalt metal–organic framework based on layered double nanosheets for enhanced electrocatalytic water oxidation in neutral media. Journal of the American Chemical Society. 2020;**142**(45):19198-19208

[51] Arrozi US, Bon V, Kutzscher C, Senkovska I, Kaskel S. Towards highly active and stable nickel-based metal–organic frameworks as ethylene oligomerization catalysts. Dalton Transactions. 2019;**48**(10):3415-3421

[52] Mousavinejad A, Rahimpour A, Shirzad Kebria MR, Khoshhal Salestan S, Sadrzadeh M, Tavajohi Hassan Kiadeh N. Nickel-Based metal–organic frameworks to improve the CO_2/CH_4 separation capability of thin-film pebax membranes. Industrial & Engineering Chemistry Research. 2020;**59**(28):12834-12844

[53] Yan J, Huang Y, Yan Y, Ding L, Liu P. High-performance electromagnetic wave absorbers based on two kinds of nickel-based MOF-derived Ni@C microspheres. ACS Applied Materials & Interfaces. 2019;**11**(43):40781-40792

[54] Liu S, Sun YY, Wu YP, Wang YJ, Pi Q, Li S, et al. Common strategy: Mounting the rod-like ni-based mof on hydrangea-shaped nickel hydroxide for superior electrocatalytic methanol oxidation reaction. ACS Applied Materials & Interfaces. 2021;**13**(22):26472-26481

[55] Zhou Y, Mao Z, Wang W, Yang Z, Liu X. In-situ fabrication of graphene oxide hybrid Ni-based metal–organic

framework (Ni–MOFs@ GO) with ultrahigh capacitance as electrochemical pseudocapacitor materials. ACS Applied Materials & Interfaces. 2016;**8**(42):28904-28916

[56] Yuan M, Wang R, Sun Z, Lin L, Yang H, Li H, et al. Morphology-controlled synthesis of Ni-MOFs with highly enhanced electrocatalytic performance for urea oxidation. Inorganic Chemistry. 2019;**58**(17):11449-11457

[57] Wang Z, Ge L, Feng D, Jiang Z, Wang H, Li M, et al. Crystal facet engineering of copper-based metal–organic frameworks with inorganic modulators. Crystal Growth & Design. 2021;**21**(2):926-934

[58] Xu WQ, He S, Lin CC, Qiu YX, Liu XJ, Jiang T, et al. A copper based metal-organic framework: Synthesis, modification and VOCs adsorption. Inorganic Chemistry Communications. 2018;**92**:1-4

[59] Ming F, Hou J, Huo D, Zhou J, Yang M, Shen C, et al. Copper-based metal–organic framework nanoparticles for sensitive fluorescence detection of ferric ions. Analytical Methods. 2019;**11**(34):4382-4389

[60] Wu Y, Ma Y, Xu G, Wei F, Ma Y, Song Q, et al. Metal-organic framework coated Fe_3O_4 magnetic nanoparticles with peroxidase-like activity for colorimetric sensing of cholesterol. Sensors and Actuators B: Chemical. 2017;**249**:195-202

[61] Chen Y, Wu H, Liu Z, Sun X, Xia Q, Li Z. Liquid-assisted mechanochemical synthesis of copper based MOF-505 for the separation of CO_2 over CH4 or N2. Industrial & Engineering Chemistry Research. 2018;**57**(2):703-709

[62] Kurisingal JF, Rachuri Y, Gu Y, Chitumalla RK, Vuppala S, Jang J, et al. Facile green synthesis of new copper-based metal–organic frameworks: Experimental and theoretical study of the co_2 fixation reaction. ACS Sustainable Chemistry & Engineering. 2020;**8**(29):10822-10832

[63] Nazari Z, Taher MA, Fazelirad H. A Zn based metal organic framework nanocomposite: Synthesis, characterization and application for preconcentration of cadmium prior to its determination by FAAS. RSC Advances. 2017;**7**(71):44890-44895

[64] Gai S, Zhang J, Fan R, Xing K, Chen W, Zhu K, et al. Highly stable zinc-based metal–organic frameworks and corresponding flexible composites for removal and detection of antibiotics in water. ACS Applied Materials & Interfaces. 2020;**12**(7):8650-8662

[65] Wang F, Xu K, Jiang Z, Yan T, Wang C, Pu Y, et al. A multifunctional zinc-based metal-organic framework for sensing and photocatalytic applications. Journal of Luminescence. 2018;**194**:22-28

[66] Park J, Hinckley AC, Huang Z, Chen G, Yakovenko AA, Zou X, et al. High thermopower in a Zn-based 3D semiconductive metal–organic framework. Journal of the American Chemical Society. 2020;**142**(49):20531-20535

[67] Shukla AK, Alam J, Alhoshan MS, Ali FAA, Mishra U, Hamid AA. Thin-film nanocomposite membrane incorporated with porous Zn-based metal–organic frameworks: Toward enhancement of desalination performance and chlorine resistance. ACS Applied Materials & Interfaces. 2021;**13**(24):28818-28831

[68] Gai S, Fan R, Zhang J, Sun J, Li P, Geng Z, et al. Structural design of low toxicity metal–organic frameworks for multifunction detection of organic and inorganic contaminants

from water. Inorganic Chemistry. 2021;**60**(14):10387-10397

[69] Safaei M, Foroughi MM, Ebrahimpoor N, Jahani S, Omidi A, Khatami M. A review on metal-organic frameworks: Synthesis and applications. TrAC Trends in Analytical Chemistry. 2019;**118**:401-425

[70] Kobielska PA, Howarth AJ, Farha OK, Nayak S. Metal–organic frameworks for heavy metal removal from water. Coordination Chemistry Reviews. 2018;**358**:92-107

[71] Awad FS, AbouZeid KM, El-MaatyWMA,El-WakilAM,El-ShallMS. Efficient removal of heavy metals from polluted water with high selectivity for mercury (II) by 2-imino-4-thiobiuret–partially reduced graphene oxide (IT-PRGO). ACS Applied Materials & Interfaces. 2017;**9**(39):34230-34242

[72] Shayegan H, Ali GA, Safarifard V. Amide-functionalized metal–organic framework for high efficiency and fast removal of Pb (II) from aqueous solution. Journal of Inorganic and Organometallic Polymers and Materials. 2020;**30**(8):3170-3178

[73] Rouhani F, Rafizadeh-Masuleh F, Morsali A. Highly electroconductive metal–organic framework: Tunable by metal ion sorption quantity. Journal of the American Chemical Society. 2019;**141**(28):11173-11182

[74] Li L, Shi Z, Zhu H, Hong W, Xie F, Sun K. Adsorption of azo dyes from aqueous solution by the hybrid MOFs/GO. Water Science and Technology. 2016;**73**(7):1728-1737

[75] Rojas S, Horcajada P. Metal–organic frameworks for the removal of emerging organic contaminants in water. Chemical Reviews. 2020;**120**(16):8378-8415

[76] Dias EM, Petit C. Towards the use of metal-organic Frameworks for water reuse: A review of the recent advances in the field of organic pollutants removal and degradation and the next steps in the field. Journal of Materials Chemistry A. 2015;**3**:22484

[77] Jin E, Kim J, Nam J, Yang DC, Jeong H, Kim S, et al. Adsorptive removal of industrial dye by nanoporous Zr porphyrinic metal–organic framework microcubes. ACS Applied Nano Materials. 2021;**4**(10):10068-10076

[78] Zhu B, Zou R, Xu Q. Metal–organic framework based catalysts for hydrogen evolution. Advanced Energy Materials. 2018:1801193

[79] Alfonso-Herrera LA, Huerta-Flores AM, Torres Martínez LM, Ramírez-Herrera DJ, Rivera-Villanueva JM. M-008: A stable and reusable metalorganic framework with high crystallinity applied in the photocatalytic hydrogen evolution and the degradation of methyl orange. Journal of Photochemistry & Photobiology A: Chemistry. 2020;**389**:112240

[80] Liu T, Li P, Yao N, Cheng G, Chen S, Luo W, et al. CoP-doped MOF-based electrocatalyst for pH-universal hydrogen evolution reaction. Angewandte Chemie. 2019;**58**(14):4679

[81] Alfonso Herrera LA, Camarillo Reyes PK, Huerta Flores AM, Martínez LT, Rivera Villanueva JM. BDC-Zn MOF sensitization by MO/MB adsorption for photocatalytic hydrogen evolution under solar light. Materials Science in Semiconductor Processing. 2020;**109**:104950

[82] Ma X, Liu F, Helian Y, Li C, Wu Z, Li H, et al. Current application of MOFs based heterogeneous catalysts in catalyzing transesterification/esterification for biodiesel production:

A review. Energy Conversion and Management. 2021;**229**:113760

[83] Cirujano FG, Dhakshinamoorthy A. Engineering of active sites in metal–organic frameworks for biodiesel production. Advanced Sustainable Systems. 2021:2100101

[84] Li Q, Chen Y, Bai S, Shao X, Jiang L, Li Q. Immobilized lipase in bio-based metal-organic frameworks constructed by biomimetic mineralization: A sustainable biocatalyst for biodiesel synthesis. Colloids and Surfaces B: Biointerfaces. 2020;**188**:110812

[85] Xie W, Wan F. Guanidine post-functionalized crystalline ZIF-90 frameworks as a promising recyclable catalyst for the production of biodiesel via soybean oil transesterification. Energy Conversion and Management. 2019;**198**:111922

[86] Peña-Rodríguez R, Márquez-López E, Guerrero A, Chiñas LE, Hernández-González DF, Rivera JM. Hydrothermal synthesis of cobalt (II) 3D metal-organic framework acid catalyst applied in the transesterification process of vegetable oil. Materials Letters. 2018;**217**:117

[87] Zhang Q, Lei D, Luo Q, Wang J, Deng T, Zhang Y, et al. Efficient biodiesel production from oleic acid using metal-organic framework encapsulated Zr-doped polyoxometalate nano-hybrids. RSC Advances. 2020;**10**:8766

Chapter 5

Adsorption and Its Applications: Using Zinc Adsorption on Water Hyacinth to Elaborate the Kinetics and Thermodynamics of Adsorption

Joseph Munene Mwaniki

Abstract

Adsorption is a technique for removing adsorbate from the liquid or gas phase using adsorbents. The adsorbent is solid while the adsorbate can either be dissolved in liquid or gas. Adsorption has attracted the attention of many researchers because of its wide applicability in water and air purification, environment friendly, effectiveness, and ease to design as compared with the other methods. Activated carbon has been used as an effective adsorbent. However, its application is limited since it's expensive. This has necessitated research interest in other materials that are safe and economical instead of commercial activated carbon. Some of the materials that have been successfully tested include sawdust, silica gel, zeolites, clay minerals and oxides, nanomaterial, agricultural by-products, biological waste, ion exchange resins and water hyacinth, etc. Although some of these materials are effective, they are not readily available. The kinetics of adsorption is done through testing the adsorption data against standard kinetic models and the model with the best line of fit, based on the values of coefficient of determination (R^2) is selected. The adsorption process is described using isotherms such as Freundlich and Langmuir. This chapter sheds more light on adsorption, the most common adsorbents, kinetic models, isotherms, and adsorption applicability.

Keywords: adsorption, adsorbents, kinetics, mechanism, isotherms, thermodynamics

1. Introduction

Over the last decade, there has been a tremendous increase in industries worldwide. Wastewater released from the industries is contaminated by toxic substances such as heavy metals. Heavy metals are persistent, fairly soluble in water, and hence easily absorbed into living cells [1]. Heavy metal pollution poses health problems such as damage to the liver, kidney, circulatory and nervous system, dermatitis, insomnia, tumor formation, rheumatoid arthritis and respiratory cancer. Heavy metals of great concern are mercury, lead, zinc, nickel, cobalt, chromium, copper, and cadmium.

Due to these and many more adverse effects of heavy metals, on humanity and the environment, there is a need for protection and restoration of the environment through the removal of heavy metals from the industrial wastewater before being released into the environment.

Several methods have been used to remove heavy metals from industrial wastewater, which include the use of membranes such as reverse osmosis, electrochemical techniques such as electrolytic extraction and electrodialysis, ion exchange, and chemical precipitation. Although these methods are effective, they are costly in terms of infrastructure, control systems, and energy [2]. There exist cheaper methods of eliminating heavy metals from wastewater such as adsorption [3]. Adsorption is becoming a regular method for the removal of heavy metals from wastewater since its relatively less costly, energy-efficient, environmentally friendly, and inexpensive [4]. Another key advantage of adsorption is the ease of designing and operating [5]. Adsorption is very effective even when the concentration of heavy metal is as low as 1 mg/L, offering metal recovery, metal selectivity and regeneration of the adsorbent material [6].

2. Adsorption mechanism

Adsorption takes place in four definitive stages: Stage 1: The mass transfer of the adsorbate by diffusion from the bulk fluid to the solid–liquid boundary layer surrounding the adsorbent particle. Stage 2: External diffusion. The transportation of the adsorbate through diffusion within the boundary layer to the outer surface of the adsorbent. Stage 3: Internal diffusion. The transfer of the adsorbate from the outer surface of the adsorbent to its inner surface by diffusion. Stage 4: The adsorption of the adsorbate on the active sites by physical (Physisorption) or chemical adsorption (Chemisorption). Physisorption is the interaction between the adsorbate and the active site through the weak van der Waals forces while in the chemical adsorption, the interaction results in the formation of a strong chemical bond. Physisorption is reversible and leads to the formation of multimolecular layers while chemisorptions are irreversible and form a unimolecular layer.

3. Common adsorbents

The applicability of the adsorbent in the removal of pollutants majorly depends on the adsorbent's characteristics. Some of the major characteristics include porosity, pore configuration, and the general nature of the surface of the adsorbent. The adsorption sites are spread throughout the solid. Adsorption sites' sizes are categorized into three: macropores, mesopores, and micropores. While macropores' diameter is larger than 50 nm, mesopores have a diameter that varies between 2 and 50 nm. Micropores' diameter is less than 2 nm. Many adsorbents are either naturally occurring or manufactured. Some of the common adsorbents include activated carbon, silica gel, zeolites, clay, nanoparticles, biological wastes, exchange resins, and water hyacinth.

3.1 Activated carbon

It is an organic material that forms a porous medium for adsorption. The structure of this medium is composite. The building blocks in this structure are carbon atoms.

Activated carbon can be used to remove substances such as dyes and pesticides as well as in the purification of wastewater [7]. This is because it is extremely effective in cadmium, lead, and zinc removal. The removal of pollutants such as chlorinated hydrocarbons, purification of helium, removal of phenols, removal of gas odors, and removal of nitrogen from the air have all utilized activated carbon [8].

3.2 Silica gel, zeolite, clay minerals, and oxides

Silica gel is a porous form of silica. It's mainly used to control moisture because of its high ability to absorb water vapor [9]. Also, it is used to control moisture in shoe boxes and to remove moisture from transformer oils and gases. Silica gel is widely used since it is cheap and can be regenerated [10]. Zeolite occurs naturally in the form of crystalline alumina that can be used in the adsorption of organic molecules from a gaseous phase. Zeolites have high water uptake since they have extensive surface area. Zeolite is used in industries to purify hydrogen gas and to recover carbon dioxide. In petroleum manufacturing, normal paraffin is separated from branched paraffin by the use of zeolites through adsorption. Clinoptilolite and bentonite, which are forms of zeolite, have been reported as proficient adsorbents in purification since they adsorb heavy metals [11, 12]. Clay minerals and oxides are available in nature and widely used because of their ability to adsorb many species of element. They adsorb cations such Cu^{2+}, anions such as N^{2+}, and neutral metallic species. Clay can be categorized into four: mica, smectite, kaolinite, and montmorillonite. Although they are readily available, they are less efficient in the adsorption of heavy metals when compared with zeolites.

3.3 Nanomaterial, agricultural by-products, and biological waste

Nanomaterial is a porous material whose pore diameter is less than 200 nm. These materials pose special properties such as distinctive surface and structural properties such as crystallinity and defect. They are used in processes such as ion exchange, catalysis, and separation. Nanomaterials are efficient in adsorption although they are expensive [13]. Most of the nanomaterials that have been used as an adsorbent include carbon nano-tube, activated carbon, and graphene. Agricultural by-products and biological wastes have been utilized in the adsorption of heavy metals. They are readily available, require modest processing, economical, offer selective adsorption and easy to generate. Agricultural by-products such as pecan shells, coconut shells, rice husks, cow dung, and maize cob have all been used effectively [14].

3.4 Ion exchange resins

These are organic materials (polymers) that can substitute ions within them with ions in a solution. This occurs when a solution containing ions is passed through the polymer. Resins can either be anionic or cationic. Anionic resins are negatively charged. As the solution passes through the resin, the positively charged ions in the solution are trapped since the resin is negatively charged [15]. Anionic resins are either weak or strong acids. Cationic resins are positively charged. They trap the negatively charged ions in the solution. The ion exchange resin is used in water softening to substitute Mg^{2+} and Ca^{2+} with Na^{+} converting hard water to soft water [16]. In this case, the resin is regenerated by rinsing it with a solution whose concentration of sodium ions is high. Resins are also used to purify water. In such a case, the poisonous heavy metal ions are replaced with ions such as sodium. Water containing no mineral content is purified using a resin that contains H^+ and OH^- to replace anions and cations.

3.5 Water hyacinth and other low-cost adsorbents

The plant can be grown in contaminated soils to absorb the heavy metal ions [17]. Also, it can be harvested, dried and ground into powder before dispensing it in heavy metal contaminated water. Water hyacinth powder should be allowed sufficient time in the contaminated water for adsorption to occur before the water can be used. The constituents of water hyacinth are cellulose (30–50%), hemicelluloses (20–40%), and lignin (15–30%) [18]. The cellulose contains functional groups such as O-H, which are involved in adsorption through deprotonation [19]. A comparative study on the efficiency of water hyacinth, water lettuce, and vetiver grass showed that the three plants have different abilities to eliminate water contaminants and their capacities are influenced by factors such as climate and temperature etc. [20]. Low-cost adsorbents provide cheap and readily available material to use as adsorbents. Lately, scientists have intensified research on the use of low-cost adsorbents in the removal of heavy metals from wastewater. Biological resources such as the agricultural waste have been put to the test and have proved useful. Rice husks, cow dung, sugarcane bagasse, sawdust, cashew nutshell, soybean hull, coconut shell, cotton hull, orange and banana peel have all successfully been used to remove heavy metals from wastewater. These adsorbents are cleaned and ground to the desired particle sizes before being used for adsorption while others are modified using modification techniques to improve the active sites [21].

4. Adsorption kinetics

During adsorption, equilibrium is not established immediately. The particle transfer from the solution to the adsorption site is restricted by the mass transfer resistance. The graph showing rate of adsorption with time is referred to as adsorption kinetics. It shows the rate of withholding or discharge of adsorbate from the aqueous solution to the adsorbent surface. The rate is influenced by the amount of the adsorbent, temperature, pH and particle size of the adsorbent, among others. The adsorption rate is restricted by resistance to diffusion in the solution (as adsorbate diffuses from the solution to the surface of the adsorbent) and within the adsorbent layers (as the adsorbate diffuses to the active site within the adsorbent) [22]. Kinetics studies are, therefore, important in determining the rate-limiting stage. Kinetics of adsorption is studied using kinetic models such as pseudo-first-order and pseudo-second-order.

4.1 Pseudo-first-order

To be able to analyze adsorption, several models have been applied. One such model is the pseudo-first-order. The linear form of the pseudo-first-order equation is [23]

$$\text{Log}\ (q_e - q_t) = \log\ (q_e) - \frac{K1}{2.303}t \quad (1)$$

Where q_e (mg/g)—the amount of adsorbate on adsorbent at equilibrium, q_t (mg/g)—the amount of heavy metal adsorbed on water hyacinth at a time t while $K_1(Min^{-1})$ is a rate constant of adsorption for pseudo-first-order. The values of q_e and q_t are determined using the adsorption capacity Eqs. (2) and (3) respectively [24].

$$(q_e) = \frac{(C1 - Ce)}{M} \times V \tag{2}$$

$$(q_t) = \frac{(C1 - C2)}{M} \times V \tag{3}$$

Where C_1 and C_2 refer to the initial and final concentration, C_e is the concentration of the analyte at equilibrium, M-mass of the adsorbent, V is the volume used. The adsorption rate constant (K_1) for pseudo-first-order is computed from the gradient of a linear plot of log (q_e-q_t) versus t while q_e is the value at the intercept.

4.2 Pseudo-second-order

The linear form of pseudo-second-order is [25];

$$\frac{t}{qt} = \frac{1}{qe}t + \frac{1}{K2qe2} \tag{4}$$

Where q_e (mg/g)—the amount of adsorbate on adsorbent at equilibrium, q_t (mg/g)—the amount of heavy metal adsorbed on water hyacinth at a time t while $K_2(Min^{-1})$ is a rate constant of adsorption for pseudo-second-order. The values of q_t are determined using the adsorption capacity Eq. (2). The adsorption rate constant for pseudo-second-order (K_2) and q_e are computed from the gradient and intercept of a linear plot of t/q_t versus t respectively. The linear plot with the highest R^2 value is considered to describe the reaction best and is taken as the correct reaction order.

4.3 Experimental example

Adsorption studies of water hyacinth powder in the removal of Zn^{2+} (95.5 ppm) from an aqueous solution were conducted, and the data obtained analyzed using both pseudo-first and pseudo-second-order.

To prepare the stock solution, the following procedure was followed [26]:

- The amount of salt dissolved in a liter of distilled water to make 1000 ppm aqueous solutions of zinc was calculated using Eq. (5) [27].

$$m = \frac{Mw}{Aw} x \frac{100}{P} x \frac{V}{1000} \tag{5}$$

Where m = mass (g) of analytical grade zinc nitrate, which was weighed, M_W = molecular weight, A_w = Atomic mass, V = volume of the stock solution to be made, P = percentage purity of the salt.

$$m = \frac{297.48}{65.4} x \frac{100}{95.97} x \frac{1000}{1000} = 4.7396g$$

- 4.7396 g weighed was transferred into a 1000 ml volumetric flask. 100 ml of distilled water was added, stirred, and made to the mark.

- Working solutions of 95.5 ppm were prepared via serial dilution following the formula indicated in Eq. (6) [28]

$$C_1V_1 = C_2V_2 \tag{6}$$

Water hyacinth stems were collected from the shores of Lake Victoria, cut into smaller pieces, and cleaned vigorously with water to get rid of dust and other contaminants. To further prepare water hyacinth pieces for adsorption, the following procedure was followed

- Thorough cleaning of water hyacinth stems using distilled water followed by sufficient sun drying. Further, oven drying at 110°C to ensure that the moisture was removed totally.

- The dried components were ground and sieved using 300 μm sieves.

4.3.1 Batch experiments

To carry out the batch studies, the following procedure was followed [29]:

- 0.5 g of groundwater hyacinth whose particle size was <300 μm was added in separate 500 ml beakers. To the beakers, a 100 ml aqueous solution of 95.5 ppm for zinc was added. The pH value of the solutions was set at 5 using a benchtop pH meter model HANNA HI 22091 and was adjusted by adding 0.1 M HNO_3 and 0.1 M NaOH.

- The mixtures were stirred continuously using a magnetic stirrer at 300 rpm for 5 min. Adsorption was allowed to occur for 2 h.

- The mixture was gravity filtered using Whatman filter paper no. 40. The residual concentrations (C_2) of selected zinc ions in the filtrate were determined using the AAS (AA-6300 SHIMADZU AAS) after 10 min for 1 h.

Table 1 gives information about the initial and final analyte concentration (C_O, and C_t), the volume of the solution (V), the mass of the adsorbent (M), the amount of adsorbate on water hyacinth at equilibrium (q_e - mg/g) and at time t (q_t-mg/g), the quotient t/q_t, the difference between q_e-q_t and their logarithmic difference (log q_e-q_t).

Time and logarithmic differences presented between in **Table 1** were used to generate **Figure 1**. **Figure 1** shows the variation of the logarithmic difference (log q_e-q_t) with time. The logarithmic difference (log qe-qt) in adsorption capacity at equilibrium and at particular time t decreased linearly as the adsorption time increased as shown between **Figure 1**.

Time and the quotient t/q_t presented in **Table 1** were used to generate **Figure 2**. **Figure 2** demonstrates how the quotient t/q_t changes with time t. The quotient of t /qt increased linearly with the adsorption time as shown in **Figure 2**.

Kinetic data for Zn^{2+}(95.5 ppm)									
time	C_t	C_o	V	M	q_t	q_e	q_e-q_t	log (q_e-q_t)	t/q_t
0	95.5	95.5	0.1	0.5	0.0000	15.6800	15.6800	1.1953	Error
10	64.1	95.5	0.1	0.5	6.2800	15.6800	9.4000	0.9731	1.5924
20	49.4	95.5	0.1	0.5	9.2200	15.6800	6.4600	0.8102	2.1692
30	42.2	95.5	0.1	0.5	10.6600	15.6800	5.0200	0.7007	2.8143
40	34	95.5	0.1	0.5	12.3000	15.6800	3.3800	0.5289	3.2520
50	21.9	95.5	0.1	0.5	14.7200	15.6800	0.9600	−0.0177	3.3967
60	17.1	95.5	0.1	0.5	15.6800	15.6800	0.0000	Error	3.8265

Table 1.
Kinetic data for Zn^{2+}.

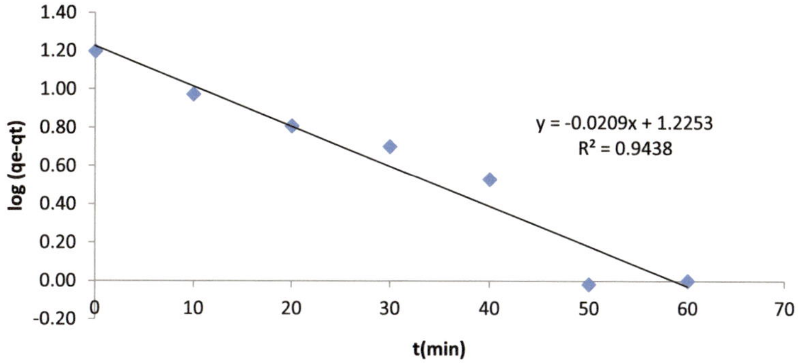

Figure 1.
Pseudo-first-order for the adsorption of 95.5 ppm zinc ions onto water hyacinth powder.

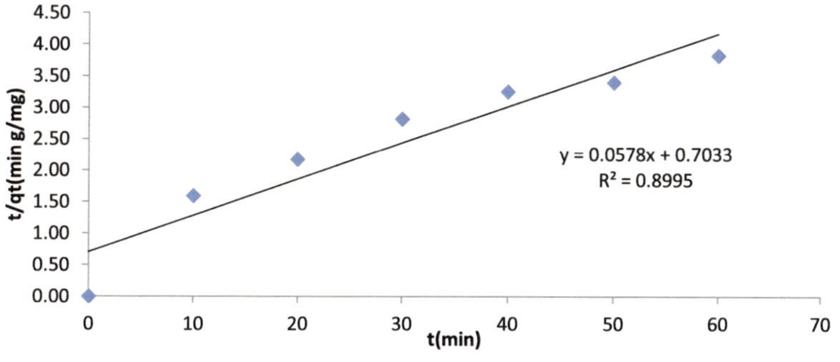

Figure 2.
Pseudo-second-order for the adsorption of 95.5 ppm zinc ions onto water hyacinth powder.

The essential information presented in **Figures 1** and **2** was used in determining the kinetic parameters, which were used in establishing the most suitable model that could be used to describe the adsorption of Zn^{2+} on water hyacinth powder.

4.3.2 Pseudo-first-order and pseudo-second-order kinetic parameters

The pseudo-first-order and pseudo-second-order kinetic parameters include K_1, q_e, and K_2. The initial concentration of Zn^{2+} was 95.5 ppm while the final

	Pseudo –first-order			Pseudo –second-order		
Metal ions	**$K_1(min^{-1})$**	**qe (mg/g)**	**R^2**	**K_2(g/mg.min)**	**qe (mg/g)**	**R^2**
Zinc ions	0.0481	16.7996	0.944	0.0048	17.3010	0.8995

Table 2.
The pseudo-first-order and pseudo second-order kinetic parameters of adsorption of zinc ions on powdered water hyacinth.

concentration was 17.1 ppm. Using the Eqs. (1), (4) and the slopes in **Figures 1** and **2**, the pseudo-first-order and pseudo-second-order kinetic parameters can be calculated as shown:

a. For the pseudo-first-order

Comparing the slope from **Figure 1** with the linear form of pseudo-first-order equation (Eq. (1))

$$y = -0.0209x + 1.2253$$

$$\text{Log}\left(q_e - q_t\right) = -\frac{K1}{2.303}t + \log\left(q_e\right)$$

$$q_e = 16.7996, \text{while } k_1 = -2.303 \text{ x} - 0.0209 = 0.0481$$

b. For the pseudo-second-order

Comparing the equation from **Figure 2** with the linear form of pseudo-second-order

$$y = 0.0578x + 0.7033$$

$$\frac{t}{qt} = \frac{1}{qe}t + \frac{1}{K2qe2.}$$

q_e = 1/0.0578, hence q_e 17.3010, K_2 = 0.0048. The pseudo-first-order and pseudo-second-order kinetic parameters are presented in **Table 2** which provides information about K_1, qe, R^2, and K_2.

The adsorption kinetics pointed out that the pseudo-first-order kinetic model expressed better the adsorption technique of zinc on water hyacinth with the regression coefficient R^2 = 0.944 being higher than that of the pseudo-second-order (R^2 = 0.8995).

5. Adsorption isotherms

Adsorption isotherm refers to a graph showing a relationship between adsorbate in the bulk and that on the surface of the adsorbent at a constant temperature. The adsorbate adsorption or release in an aqueous solution could be represented as follows [30]:

$$S_f(Me) + H^+ \text{ (solution)} \leftrightarrow S_f \text{ (adsorbent)} + Me(\text{solution}) \tag{7}$$

where S_f represents the different adsorption sites on the adsorbent where the metal, Me can be retained. The adsorption of the metal on the sites could be enhanced by both the physical and chemical characteristics of the medium [31]. Langmuir and Freundlich's isotherms have been applied to heavy metal adsorption studies. The Langmuir isotherm assumes that adsorption is a reversible process and the adsorbing material has a definite number of active spots, which are evenly distributed [32]. The Freundlich isotherm describes multilayer adsorption [33]. The adsorption isotherms and modeling are very important in determining the accuracy of the adsorption process.

5.1 Langmuir isotherms

Langmuir adsorption was intended to illustrate gas–solid adsorption. However, it is also used to compute and compare the adsorption capacities of various adsorbents. Langmuir isotherm balances both adsorption and desorption rates. Adsorption is the measure of the portion of the adsorbent's surface that is open while desorption accounts for the portion of the surface of the adsorbent that is occupied [33]. The linear form of Langmuir isotherm that was used is shown in Eq. (8) [34]

$$\frac{1}{qe} = \left(\frac{1}{qmb}\right)\cdot\frac{1}{Ce} + \frac{1}{qm} \tag{8}$$

where: q_e (mg/g)—equilibrium adsorption capacity, C_e (mg/l)—the amount of adsorbed adsorbate at equilibrium, q_m (mg/g)—the highest amount of the adsorbate for every unit weight of adsorbent while b (l/mg)—Langmuir constant (binding affinity). The q_m and b values are determined graphically from a plot of $1/q_e$ against $1/C_e$. The dimensionless factor R_L, also known as the separation factor, is used to describe the Langmuir isotherm. It is determined using Eq. (9) [35]

$$R_L = \frac{1}{b\ Co + 1} \tag{9}$$

where; $R_L < 1$—adsorption is favorable,
$R_L > 1$—adsorption is unfavorable
$R_L = 1$—linear
$R_L > 1$—adsorption is irreversible

5.2 Freundlich isotherms

Freundlich isotherm equation is based on the fact that adsorption occurs on a heterogeneous surface. It is an experiential model, which takes into account the adsorptive active sites and their energy exponentially. The expression also considers the heterogeneity of the adsorbent's surface. It is expressed as shown in Eq. (10) [36]

$$q_e = K_F {C_e}^{1/n} \tag{10}$$

The linear form of the Freundlich isotherm equation is shown in Eq. (11) [37].

$$log\ q_e = \frac{1}{n}\ log\ C_e + log\ K_f \tag{11}$$

Where qe (mg/g)—the amount of the adsorbate adsorbed per unit weight of water hyacinth bio-material, Ce (mg/L)—the amount of unadsorbed adsorbent in the solution, K_f—a constant indicating adsorption capacity while n—adsorption intensity. Adsorption studies of zinc ions conducted using synthesized magnetite and baobab composite showed that the equilibrium data were suitably expressed by Freundlich because of the high correlation coefficient [38].

5.3 Experimental example

Equilibrium adsorption studies of water hyacinth powder in the removal of Zn^{2+} from aqueous solution were performed, and the data obtained were analyzed using both Langmuir and Freundlich isotherms. The working solutions of concentrations 0.5, 1.0, 10, 20, 40, 50, 60, 80, 100 ppm were prepared through serial dilution of the stock solution prepared in Section 4.3 following the formula indicated in Eq. (6).

5.3.1 Batch experiments

100 ml of aqueous solutions containing zinc metal ion (0.5–100 ppm) were added to 0.5 g of water hyacinth powder (<300 μm) in separate 500 ml beakers. The pH of the solutions was set at 5, and the solutions were stirred for 5 minutes. Adsorption was allowed to occur for 120 minutes. These experiments were performed at room temperature until the equilibrium was established. The mixture was filtered using gravity filtration (using Whatman Filter Paper no. 40 and plastic filter funnels). The residual concentrations of zinc ions in the filtrate were determined using the Atomic Absorption Spectrometer (AA-6300 SHIMADZU AAS). The data obtained from the adsorption studies were computed using Eqs. (2), (8) and (11), and tabulated in **Table 3**. **Table 3** gives information about the initial and equilibrium analyte concentration (C_0, and C_e), reciprocal of equilibrium concentration ($1/C_e$), the amount of zinc ion on water hyacinth at equilibrium (q_e - mg/g) and its reciprocal ($1/q_e$), and the logarithmic values of both q_e and C_e (log q_e, log C_e).

The logarithmic values of C_e and q_e accessible in **Table 3** were used to generate **Figure 3** which demonstrates demonstrates the variation of log q_e with log C_e.

Equilibrium data for Zn^{2+}						
C_o	C_e	$1/C_e$	q_e	$1/q_e$	log q_e	log C_e
0.5	0	Error	0.1000	10.0000	Error	−1.0000
1	0	Error	0.2000	5.0000	Error	−0.6990
10	0	Error	2.0000	0.5000	Error	0.3010
20	2.4	0.4167	3.5200	0.2841	0.3802	0.5465
40	4.9	0.2041	7.0200	0.1425	0.6902	0.8463
50	9.4	0.1064	8.1200	0.1232	0.9731	0.9096
60	11.1	0.0901	9.7800	0.1022	1.0453	0.9903
80	16.4	0.0610	12.7200	0.0786	1.2148	1.1045
100	20.7	0.0483	15.8600	0.0631	1.3160	1.2003

Table 3.
Equilibrium data for Zn^{2+}.

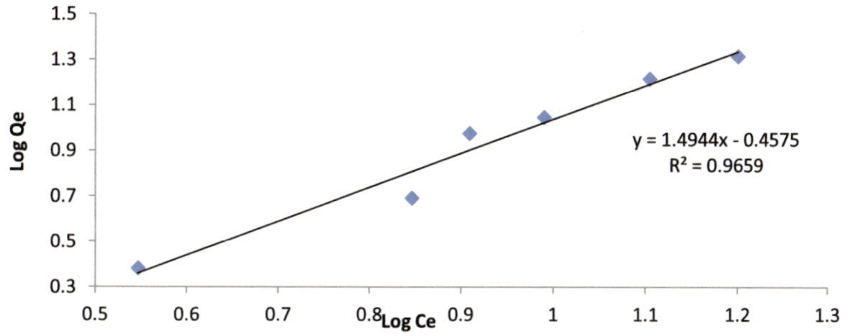

Figure 3.
Linearized Freundlich plot for the adsorption of Zinc ions onto water hyacinth powder.

The logarithm of q_e (adsorption capacity at equilibrium) increased linearly to the logarithm of C_e (concentration at equilibrium) as shown in **Figure 3**.

The quotients $1/C_e$ and $1/q_e$ presented in **Table 3** were used to generate **Figure 4** which illustrates how the quotient $1/C_e$ varies with $1/q_e$. The reciprocal of q_e (adsorption capacity at equilibrium) increased linearly with the reciprocal of C_e (concentration at equilibrium) as shown in **Figure 4**.

The information presented in **Figures 3** and **4** was used in computing the Freundlich and Langmuir parameters essential in determining the most suitable isotherm for explaining the adsorption of Zn^{2+} on water hyacinth powder.

5.3.2 Freundlich and Langmuir parameters

The Freundlich and Langmuir parameters include q_m, b, n, and K_f. The initial concentration of Zn^{2+} was 100 ppm while the equilibrium concentration was 20.7 ppm. Using the Eqs. (8), (11) and the slopes from **Figures 3** and **4**, the Langmuir and Freundlich parameters were calculated as shown;

a. Freundlich parameters

Comparing Eq. (11) with the equation from **Figure 3**

$$y = 1.4944x - 0.4575$$

$$\log q_e = \frac{1}{n} \log C_e + \log K_f$$

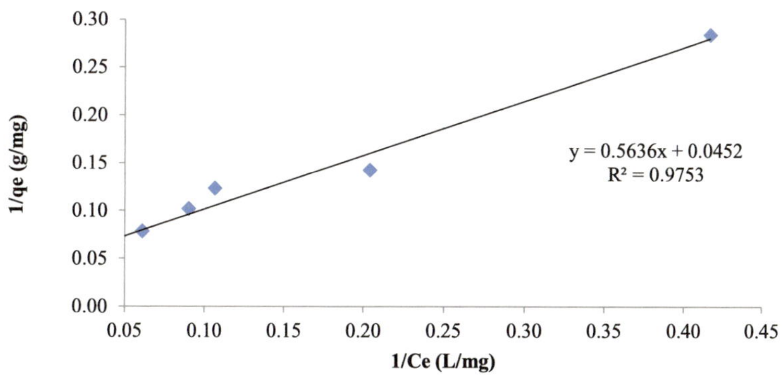

Figure 4.
Linearized Langmuir plot for the adsorption of Zinc ions onto water hyacinth powder.

Freundlich isotherm	Langmuir isotherm
n = 0.6692	b (L/mg) =0.0802
K_f = 0.3487	q_m (mg/g) =22.1239
R^2 = 0.9659	R^2 = 0.975

Table 4.
Langmuir and Freundlich isotherms parameters for adsorption of zinc ions from aqueous solution ground using water hyacinth powder.

$$log\ K_f = 0.3276 \text{ hence } K_f = 0.3487$$

$$1/n = 1.4944 \text{ hence } n = 0.6692$$

b. Langmuir parameters

Comparing Eq. (8) with equation from **Figure 4**;

$$y = 0.5636x + 0.0452$$

$$\frac{1}{qe} = \left(\frac{1}{qmb}\right) \cdot \frac{1}{Ce} + \frac{1}{qm}$$

$$\frac{1}{qm} = 0.0452 \text{ hence } q_m = 22.1239$$

$$\frac{1}{qm.b} = 0.5636 \text{ replacing } q_m \text{ we get } b = 0.0802$$

From Eq. (9), $R_L = \frac{1}{b\ Co+1}$. The value of $R_L = \frac{1}{(0.0802x\ 100)+1} = 0.1109$.

The Freundlich and Langmuir adsorption parameters are presented in **Table 4** which provides information about n, b, K_f, q_m, and the regression coefficient of **Figures 3** and **4**.

The use of water hyacinth powder in the adsorption of zinc ions correlated well with the Langmuir model in contrast to the Freundlich model since it had the highest regression coefficient (R^2 = 0.975). The linearity of **Figure 4** demonstrates the validity of the Langmuir isotherm whose basis is the formation of a monolayer on the surface of water hyacinth powder. Results similar to these were obtained by [39] where the equilibrium data during adsorption of Zinc (II) ions from aqueous solution using functionalized lignocelluloses derived from waste biomass was suitably represented by Langmuir isotherm due to high correlation coefficients. The adsorption process was favorable since the separation factor R_L (0.1109) is less than 1. The Freundlich parameter *n* (n = 0.6692) indicates that the adsorption (physical adsorption) of zinc is unfavorable since $2 < n < 10$ indicates favorable adsorption. The Langmuir constant b is important in establishing the adsorption affinity that relates the bond energy between adsorbent and adsorbate [40].

6. Thermodynamics of adsorption

To evaluate the adsorption process further, it is imperative to examine the adsorbate's binding thermodynamics on the adsorbent. Thermodynamics include changes in entropy($\Delta S°$), enthalpy ($\Delta H°$), and Gibbs free energy ($\Delta G°$), which are computed using Eqs. (12)–(14).

The Gibbs free energy$\Delta G°$is given by [41];

$$\Delta G° = -R\,T\,ln\ Kc \tag{12}$$

Where $\Delta G°$ (kJ/mol) is the change in the Gibbs free energy of adsorption, R is the gas constant (8.314 J/ (Kmol)), T is the temperature in Kelvin, K_c is the equilibrium constant. Equilibrium constant K_c can be stated with reference to Gibbs free energy $\Delta G°$and entropy $\Delta S°$ as indicated in the Van't Hoffs reaction given in Eq. (13) [42].

$$ln\ Kc = -\frac{\Delta H°}{R}\left(\frac{1}{T}\right) + \frac{\Delta S°}{R} \tag{13}$$

Where is $\Delta H°$ (kJ/mol) enthalpy change while $\Delta S°$ (J/ (mol K) is entropy change. K_c is computed by following Eq. (14) [43].

$$Kc = \frac{Qe}{Ce} \tag{14}$$

Where Qe is the equilibrium adsorption capacity while Ce is the equilibrium concentration of the heavy metal under investigation.

To determine thermodynamic parameters of adsorption, the following procedure is followed;

1. Preparation of stock solution from soluble heavy metal salt. The working solution is prepared through serial dilution of the stock solution.

2. Batch experiments studies should be carried out at a specified temperature until equilibrium is established. The procedure should be repeated with other temperatures under study (Batch studies can be carried out at temperatures such as 298,303,313,323 K). The residual adsorbate concentration at equilibrium (Ce) should be determined for each temperature. The equilibrium adsorption capacity (Qe) should then be computed using Eq. (2).

3. The values of Qe and Ce obtained for each temperature should be used to compute Kc values.

4. Further, the ln of Kc values obtained should be calculated for each temperature.

5. Using ln Kc and 1/T values obtained, a plot showing how ln Kc (y-axis) varies with 1/T (x-axis) should be drawn using excel.

6. The straight-line graph equation (y = Mx + c) obtained from the plot should be compared with Eq. (13). With the slope and R (gas constant), the values of $\Delta H°$can be calculated. With the Y-intercept (Value of C in the straight-line graph) and the R (gas constant), the values of $\Delta S°$can be computed easily

7. The Gibbs free energies for each temperature are finally calculated and results interpreted.

Gibbs free energy (ΔG°)	Enthalpy(ΔH°)	Entropy (ΔS°)	Spontaneity
ΔG <0	ΔH < 0	ΔS > 0	Spontaneous at all Temperatures
ΔG <0	ΔH >0	ΔS > 0	Spontaneous at high Temperatures
ΔG <0	ΔH <0	ΔS < 0	Spontaneous at low Temperatures
ΔG <0	ΔH > 0	ΔS <0	Non Spontaneous at all Temperatures

Table 5.
Gibbs free energy (ΔG°), enthalpy (ΔH°), entropy (ΔS°) and spontaneity of an adsorption process.

Negative values of ΔG°indicate that the process is spontaneous and will support the forward reaction while positive values show the process is nonspontaneous, hence will support the reverse reaction [44]. The spontaneity of the adsorption process depends on the values of ΔH°andΔS° as outlined in **Table 5**. **Table 5** outlines information about Gibbs free energy (ΔG°) and how the spontaneity of an adsorption process is influenced by both enthalpy (ΔH° and entropy (ΔS°).

When ΔH° >0, then the adsorption process is endothermic and requires to absorb heat from the environment. When ΔH° < 0, the adsorption process is exothermic and releases heat into the surroundings [45]. If ΔS° < 0, it indicates that the adsorption is orderly but if ΔS° > 0 shows adsorption is disorderly at the surface of the adsorbent [46].

7. Application of adsorption

1. Getting rid of colors—Juice obtained from sugar cane is treated with animal charcoal to eliminate the coloring agent to get a clear liquid.

2. Separation of inert gases—A stream of inert gases can be separated into constituent gases through the use of coconut charcoal since the noble gases are adsorbed at different degrees.

3. Controlling humidity—Silica and aluminum gels are applied as adsorbents in the removal of moisture.

4. Gas mask—In coal mines, gas masks are worn to aid staff in breathing. A gas mask consists of activated charcoal that adsorbs impurities.

5. Dyeing cloth—Mordants are applied in dyeing. The mordant, e.g., alum adsorbs the dye elements, hence the dye does not cling to the cloth material.

8. Conclusion

In the review, we have examined the adsorption process, adsorbent, adsorption mechanism, and some of the common adsorption applications. Also, we have scrutinized essential kinetic and thermodynamic parameters of adsorption. Kinetic models and isotherms described include pseudo-first-order, pseudo-second-order, Langmuir, and Freundlich. Kinetic parameters explained include adsorption capacity, pseudo-first and pseudo-second constants while the isotherm parameters described herein include adsorption intensity (n), binding affinity (b), Freundlich constant (K_f), adsorption capacity (q_m), separation factor (R_L). We have used the adsorption of zinc ions to illustrate how kinetic and isotherm parameters are computed during adsorption studies. The adsorption data for zinc from the aqueous solution using water hyacinth fitted well in the Langmuir model (R^2 = 0.975) in comparison to the Freundlich model (R^2 = 0.9659) based on the correlation coefficients. Additionally, the study showed that zinc adsorption obeys the pseudo-first-order reaction kinetic model. The adsorption of zinc ions using water hyacinth powder was favorable since the separation factor R_L was less than 1. The Freundlich parameter n an indicator of the probability of physical adsorption taking place showed that physisorption was unfavorable since $n < 2$. Further, this chapter clearly describes how thermodynamics parameters (entropy($\Delta S°$), enthalpy ($\Delta H°$), and Gibbs free energy ($\Delta G°$)) are evaluated during adsorption studies. Also, the chapter details the meaning of entropy($\Delta S°$), enthalpy ($\Delta H°$), and Gibbs free energy ($\Delta G°$) concerning the spontaneity of the adsorption process.

Acknowledgements

I am deeply obliged to my supervisors, Dr. John Onam Onyatta and Prof. Amir O. Yusuf, for the invaluable comments, suggestions, corrections, patience, and constructive criticisms that went a long way to refine this work. Without whose help, this work would not have been a success. Thank you very much for making sure I kept my focus.

Originality/declaration

The author declares that this work is their original work and has not been submitted elsewhere for examination or publication.

Conflict of interest

The author declares no conflict of interest.

Author details

Joseph Munene Mwaniki
University of Nairobi, Nairobi, Kenya

*Address all correspondence to: josemunene77@gmail.com

References

[1] Kinuthia GK, Ngure V, Beti D, et al. Levels of heavy metals in wastewater and soil samples from open drainage channels in Nairobi, Kenya: Community health implication. Scientific Reports. 2010;**10**:8434. DOI: 10.1038/s41598-020-65359-5

[2] Renu A, M., & Singh, K. Heavy metal removal from wastewater using various adsorbents: A review. Journal of Water Reuse and Desalination. 2017;7(4): 387-419. DOI: 10.2166/wrd.2016.104

[3] Qaseem NAA, Mohammed RH, Lawal DU. Removal of heavy metal ions from waste water: a comprehensive and critical review. npj Clean Water. 2021;**4**: 36. DOI: 10.1038/s41545-021-00127-0

[4] Vo TS, Hossain MM, Jeong HM, et al. Heavy metal removal applications using adsorptive membranes. Nano Convergence. 2020;7:36

[5] Moosavi S, Lai CW, Gan S, Zamiri G, Akbarzadeh Pivehzhani O, Johan MR. Application of efficient magnetic particles and activated carbon for dye removal from wastewater. ACS Omega. 2020;**5**(33):20684-20697. DOI: 10.1021/acsomega.0c01905

[6] Ugwu EI, Tursunov O, Kodirov D, Shaker LM, Al-Amiery AA, Yangibaeva I, et al. Adsorption mechanisms for heavy metal removal using low cost adsorbents: A review. IOP Conference Series: Earth and Environmental Science. 2020;**614**(1): 012166. DOI: 10.1088/1755-1315/614/1/012166

[7] Saleh IA, Zouari N, Al-Ghouti MA. Removal of pesticides from water and wastewater: Chemical, physical and biological treatment approaches. Environmental Technology & Innovation. 2020;**19**:101026. DOI: 10.1016/j.eti.2020.101026

[8] Fang M-L, Chang H-Y, Chen C-H, Lin S-L, Hsieh Y-K, Chou M-S, et al. Chemical adsorption of nitrogen dioxide with an activated carbon adsorption system. Aerosol and Air Quality Research. 2019;**19**(11):2568-2575. DOI: 10.4209/aaqr.2019.09.0439

[9] Grande CA, Morence DGB, Bouzga, Aud. M., & Andreassen, K. A. Silica gel as a selective adsorbent for biogas drying and upgrading. Industrial & Engineering Chemistry Research. 2020;**59**(21): 10142-10149. DOI: 10.1021/acs.iecr.0c00949

[10] Yan KL, Wang Q. Adsorption characteristics of the silica gels as adsorbent for gasoline vapors removal. IOP Conference Series: Earth and Environmental Science. 2018;**153**(2): 022010. DOI: 10.1088/1755-1315/153/2/022010

[11] Kumar S, Kumar P, Jasra RV. Sorption of HCl from an aromatic hydrocarbon mixture using modified molecular sieve zeolite 13X. ACS Omega. 2021;**6**(43):28742-28751. DOI: 10.1021/acsomega.1c03450

[12] Hussain T, Hussain AI, Chatha SAS, Ali A, Rizwan M, Ali S, et al. Synthesis and characterization of Na-zeolites from textile waste ash and its application for removal of Lead (Pb) from wastewater. International Journal of Environmental Research and Public Health. 2021;**18**(7): 3373. DOI: 10.3390/ijerph18073373

[13] Yang J, Hou B, Wang J, Tian B, Bi J, Wang N, et al. Nanomaterials for the removal of heavy metals from wastewater. Nanomaterials. 2019;**9**(3): 424. DOI: 10.3390/nano9030424

[14] Bansode RR, Losso JN, Marshall WE, Rao RM, Portier RJ. Adsorption of volatile organic compounds by pecan shell- and almond shell-based granular activated carbons. Bioresource Technology. 2003;**90**(2):175-184. DOI: 10.1016/S0960-8524(03)00117-2

[15] Gürkan EH, İlyas B, Tibet Y. Adsorption of Cu(II) Ve Zn(II) ions by alginate-based composites: Full factorial design approach. In: Fullerenes, Nanotubes and Carbon Nanostructures. London, UK: Taylor and Francis; 2022. pp. 1-14. DOI: 10.1080/1536383X.2021.2021891

[16] Sahin S, Dykstra JE, Zuilhof H, Zornitta RL, de Smet LCPM. Modification of cation-exchange membranes with polyelectrolyte multilayers to tune ion selectivity in capacitive deionization. ACS Applied Materials & Interfaces. 2020;**12**(31): 34746-34754. DOI: 10.1021/acsami.0c05664

[17] Jones JL, Jenkins RO, Haris PI. Extending the geographic reach of the water hyacinth plant in removal of heavy metals from a temperate northern hemisphere river. Scientific Reports. 2018;**8**(1):11071. DOI: 10.1038/s41598-018-29387-6

[18] Sanmuga Priya E, Senthamil Selvan P. Water hyacinth (Eichhornia crassipes) – An efficient and economic adsorbent for textile effluent treatment – A review. Arabian Journal of Chemistry. 2017;**10**:S3548-S3558. DOI: 10.1016/j.arabjc.2014.03.002

[19] Iftekhar S, Ramasamy DL, Srivastava V, Asif MB, Sillanpää M. Understanding the factors affecting the adsorption of lanthanum using different adsorbents: A critical review. Chemosphere. 2018;**204**:413-430. DOI: 10.1016/j.chemosphere.2018.04.053

[20] Rezania S, Md Din MF, Eva Mohamad S, Sohaili J, Mat Taib S, Mohd Yusof MB, et al. Review on pretreatment methods and ethanol production from cellulosic water hyacinth. BioResources. 2017;**12**(1):2108-2124. DOI: 10.15376/biores.12.1.Rezania

[21] Adeyemo AA, Adeoye IO, Bello OS. Adsorption of dyes using different types of clay: A review. Applied Water Science. 2017;**7**(2):543-568. DOI: 10.1007/s13201-015-0322-y

[22] Zand AD, Abyaneh MR. Adsorption of Lead, manganese, and copper onto biochar in landfill leachate: Implication of non-linear regression analysis. Sustainable Environment Research. 2020;**30**(1):18. DOI: 10.1186/s42834-020-00061-9

[23] Xiao Y, Azaiez J, Hill JM. Erroneous application of pseudo-second-order adsorption kinetics model: Ignored assumptions and spurious correlations. Industrial & Engineering Chemistry Research. 2018;**57**(7):2705-2709. DOI: 10.1021/acs.iecr.7b04724

[24] Onwordi CT, Uche CC, Ameh AE, Petrik LF. Comparative study of the adsorption capacity of lead (II) ions onto bean husk and fish scale from aqueous solution. Journal of Water Reuse and Desalination. 2019;**9**(3):249-262. DOI: 10.2166/wrd.2019.061

[25] Revellame ED, Fortela DL, Sharp W, Hernandez R, Zappi ME. Adsorption kinetic modeling using pseudo-first order and pseudo-second order rate laws: A review. Cleaner Engineering and Technology. 2020;**1**:100032. DOI: 10.1016/j.clet.2020.100032

[26] Mwaniki J, Onyatta J, Amir Y. Adsorption of heavy metal ions from aqueous solutions and wastewater using water hyacinth powder. International Journal of Trend in Scientific Research

And Development. 2019;**4**(1):2456-6470 Available from: https://www.ijtsrd.com/papers/ijtsrd29 419.pdf

[27] Burca S, Indolean C, Maicaneanu A. Malachite green dye adsorption from model aqueous solutions using corn cob activated carbon (CCAC). Studia universitatis babes-Bolyai. Chemia. 2017; **62**(4):293+ Available from: https://link.gale.com/apps/doc/A526440889/AONE?u=anon~d0e96bfd&sid=googleScholar&xid=57bfe6ed

[28] Chong SH. Wither the concepts of mole and concentration: Conceptual confusion in applying $M_1V_1 = M_2V_2$. Universal Journal of Educational Research. 2016;**4**(5):1158-1162. DOI: 10.13189/ujer.2016.040527

[29] Munene JM, Onyatta JO, Yusuf AO. Characterization of water hyacinth powder using FTIR spectroscopy and the adsorption behaviour of Pb^{2+}, Cd^{2+}, Zn^{2+}, Ni^{2+} and Cr^{2+} in aqueous solution. Asian Journal of Applied Chemistry Research. 2020;**6**:47-55. DOI: 10.9734/ajacr/2020/v6i130151

[30] Damaris M, Duke OO, Graham J, David K. Investigation of Kenyan bentonite in adsorption of some heavy metals in aqueous systems using cyclic voltammetric techniques. International Journal of Physical Sciences. 2014;**9**(5):102-108. DOI: 10.5897/IJPS2013.4095

[31] Ambaye TG, Vaccari M, van Hullebusch ED, Amrane A, Rtimi S. Mechanisms and adsorption capacities of biochar for the removal of organic and inorganic pollutants from industrial wastewater. International journal of Environmental Science and Technology. 2021;**18**(10):3273-3294. DOI: 10.1007/s13762-020-03060-w

[32] Chen Q, Tian Y, Li P, Yan C, Pang Y, Zheng L, et al. Study on shale adsorption equation based on monolayer adsorption, multilayer adsorption, and capillary condensation. Journal of Chemistry. 2017;**2017**:1-11. DOI: 10.1155/2017/1496463

[33] Ayawei N, Ebelegi AN, Wankasi D. Modelling and interpretation of adsorption isotherms. Journal of Chemistry. 2017b;**2017**:1-11. DOI: 10.1155/2017/3039817

[34] Swenson H, Stadie NP. Langmuir's theory of adsorption: A centennial review. Langmuir. 2019;**35**(16):5409-5426. DOI: 10.1021/acs.langmuir.9b00154

[35] Akrawi HSY, Al-Obaidi MA, Abdulrahman CH. Evaluation of Langmuir and Frendlich isotherm equation for zinc adsorption in some calcareous soil of Erbil province north of Iraq. IOP Conference Series: Earth and Environmental Science. 2021;**761**(1):012017. DOI: 10.1088/1755-1315/761/1/012017

[36] Nnaji CC, Agim AE, Mama CN, Emenike PC, Ogarekpe NM. Equilibrium and thermodynamic investigation of biosorption of nickel from water by activated carbon made from palm kernel chaff. Scientific Reports. 2021;**11**(1):7808. DOI: 10.1038/s41598-021-86932-6

[37] Palanivell P, Ahmed OH, Latifah O, Abdul Majid NM. Adsorption and desorption of nitrogen, phosphorus, potassium, and soil buffering capacity following application of chicken litter biochar to an acid soil. Applied Sciences. 2019;**10**(1):295. DOI: 10.3390/app10010295

[38] Abdus-Salam N, Adekola SK. Adsorption studies of zinc(II) on magnetite, baobab (Adansonia digitata) and magnetite–baobab composite. Applied Water Science. 2018;**8**(8):222. DOI: 10.1007/s13201-018-0867-7

[39] Wadhawan S, Jain A, Nayyar J, Mehta SK. Role of nanomaterials as adsorbents in heavy metal ion removal from waste water: A review. Journal of Water Process Engineering. 2020;**33**: 101038. DOI: 10.1016/j.jwpe.2019.101038

[40] Togue Kamga F. Modeling adsorption mechanism of paraquat onto Ayous (Triplochiton scleroxylon) wood sawdust. Applied Water Science. 2019; **9**(1):1. DOI: 10.1007/s13201-018-0879-3

[41] Dang J, Wang H, Wang C. Adsorption of toxic zinc by functionalized lignocellulose derived from waste biomass: Kinetics, isotherms and thermodynamics. Sustainability. 2021;**13**(19):10673. DOI: 10.3390/su131910673

[42] Ebelegi AN, Ayawei N, Wankasi D. Interpretation of adsorption thermodynamics and kinetics. Open Journal of Physical Chemistry. 2020; **10**(03):166-182. DOI: 10.4236/ojpc.2020.103010

[43] Khayyun TS, Mseer AH. Comparison of the experimental results with the Langmuir and Freundlich models for copper removal on limestone adsorbent. Applied Water Science. 2019;**9**(8):170. DOI: 10.1007/s13201-019-1061-2

[44] Fendi JW, Naser AJ. Adsorption isotherms study of methylene blue dye on membranes from electrospun nanofibers. Oriental Journal of Chemistry. 2018;**34**(6):2884-2894. DOI: 10.13005/ojc/340628

[45] Al-Ghouti MA, Al-Absi RS. Mechanistic understanding of the adsorption and thermodynamic aspects of cationic methylene blue dye onto cellulosic olive stones biomass from wastewater. Scientific Reports. 2020;**10**(1):15928. DOI: 10.1038/s41598-020-72996-3

[46] Edet UA, Ifelebuegu AO. Kinetics, isotherms, and thermodynamic modeling of the adsorption of phosphates from model wastewater using recycled brick waste. PRO. 2020; **8**(6):665. DOI: 10.3390/pr8060665

Chapter 6

Adsorption of Chromium from an Aqueous Solution onto Chitosan Beads Modified with Sodium Dodecyl Sulfate (SDS)

Naoki Kano, Zou Ming, David Eva Vanessa Anak and Muhammad Nabil Md Sari

Abstract

The goal of this research is to make chitosan beads that have been treated with sodium dodecyl sulfate (SDS) to remove chromium (Cr) from an aqueous solution effectively. The successful synthesis of the SDS-chitosan was proven through characterization, which were carried out using by scanning electron microscopy–energy dispersive X–ray spectroscopy (SEM-EDS), Fourier transform-infrared spectroscopy (FT-IR) and X-ray photoelectron spectroscopy (XPS). The adsorption of Cr on the SDS material was investigated by varying experimental conditions such as pH, contact time and adsorbent dosage. The maximum adsorption capacity of SDS-chitosan for Cr (III) was estimated to be 3.42 $mg{\cdot}g^{-1}$ and 3.23 $mg{\cdot}g^{-1}$ for Cr(VI). Based on the results of adsorption kinetics and isothermal models, the adsorption process conform to the pseudo-second-order and Langmuir isotherm models. This indicates that the adsorption of Cr on SDS-chitosan is mainly dominated by chemical adsorption and monolayer reaction. In addition, according to thermodynamic analyses, the adsorption of Cr is an endothermic reaction. These results show that the new adsorbent has obvious application prospect for removing Cr.

Keywords: chitosan beads, sodium dodecyl sulfate (SDS), chromium, adsorption, adsorption isotherms, adsorption kinetics

1. Introduction

Due to the rapidly growing number of manufacturing industries, toxic metal contamination in aquatic environments has gotten a lot of attention. Among the contaminants, heavy metals are targeted for major environmental concern because they are non-biodegradable, and they cannot be decomposed or metabolized [1]. Several metals cause serious health and environmental problems, and chromium (Cr) compounds are one of the most toxic contaminants in wastewater due to their high solubility and toxicity, as well as their free transferability [2].

Cr has been widely applied in a variety of industrial activities due to its excellent properties, including electroplating, leather tanning, nuclear power plants, and textile industries [3, 4]. Furthermore, it can be used for anodizing, corrosion control, and chemical manufacturing [5–7]. In a natural environment, Cr usually exists in two stable oxidation states: trivalent Cr(III) and hexavalent Cr(VI). Meanwhile, other oxidation states are not stable in aerated aqueous media [8]. Specifically for Cr(VI), it may exist in the form of CrO_4^{2-} or $HCrO_4^{-}$ in a natural aqueous environment, whereas Cr(III) is inclined to form $[Cr(H_2O)_6]^{3+}$, $Cr(H_2O)_5(OH)^{2+}$, $Cr(H_2O)_4(OH)^{2+}$, or Cr(III) organic complexes.

Given its considerable risk to biological systems, many studies have focused on the removal of Cr(VI). Cr(VI) is highly toxic, carcinogenic, and mutagenic [8, 9]. Adverse health effects have been linked to Cr(VI) exposure, such as bronchitis, liver damage, kidney damage, brain damage, and even lung cancer. On the other hand, Cr(III) is the most stable form in reducing conditions and it exists as cationic species $Cr(OH)_2^{+}$ and $Cr(OH)_2^{+}$, with the first or second hydrolysis products dominating at pH values ranging from 4 to 8. Although Cr(III) is an essential microelement for the effective maintenance of mammal's glucose, lipid, and protein metabolism [10], high doses of Cr(III) may cause negative consequences to the environment. Moreover, there are currently just a few articles on the adsorption of Cr(III). Therefore, the development of a recovery method for this metal (both Cr(III) and Cr(VI)) is significant from an environmental aspect.

Ion exchange, precipitation, ultrafiltration, reverse osmosis, and electro dialysis are one of the physical and chemical technologies that have been reported for the removal of heavy metals [11]. However, these procedures have some drawbacks, such as a high consumption of reagents and energy, low selectivity, high operational costs, and difficult further treatment due to toxic sludge production [12]. Adsorption is an effective method for removing metallic ions from aqueous solutions [10, 13]; and biological adsorption (biosorption) is one of the most environmentally friendly, cost-effective, recyclable, and technically simple methods [14, 15].

Among the many biosorbents available, chitosan can be an excellent biosorbent for metals because its amine ($-NH_2$) and hydroxyl (-OH) groups may serve as coordination sites to form complexes with various heavy metal ions [16]. Chitosan, whose full chemical name is (1,4)-2-amino-2- deoxy-β-D-glucose, has been proven to be particularly effective as a biosorbent for the recovery of several toxic metals, including mercury (Hg), uranium (U), molybdenum (Mo), vanadium (V), and platinum (Pt) [17–19]. It can be employed as an environmentally friendly adsorbent because it is cost effective, and it does not result in secondary pollution. Chitosan is a polymer that is made via alkaline deacetylation of chitin, which comes from cellulose, the most abundant biopolymer. It can be acquired from the shells of seafood, such as prawns, crabs, and lobsters [20]. The biopolymer has a high nitrogen content, which is present in the form of amine groups, free amino groups, and hydroxyl groups, all of which are responsible for metal ion binding through chelation mechanisms [21].

Despite the uses of chitosan in the removal of various pollutants have been adequately reviewed [22], on the other hand, it has some defects, including notable swelling in aqueous media and nonporous structures, resulting in a very low surface area [23]. Therefore, a variety of chemical modifications can be used to produce chitosan derivatives that improve the removal efficiency of the heavy metal [24]. For example, to increase the number of exposed active sites, several chemical or physical modifications can be adopted [25, 26]. Moreover, since silicon dioxide has numerous properties, including rigid configuration, porosity, and large surface area, it can be employed to counteract the defects of chitosan. In addition, modified silicon dioxide has been produced through a graft between silanol groups and ligands [27–29]. In prior work, we synthesized a hybrid membrane of carboxymethyl chitosan and silicon dioxide as adsorbents for the removal of Cr(VI) [30]. Furthermore, we used epichlorohydrin (EP) and glutaraldehyde (GA) as cross-linked agents in an adsorption experiment of chromate ions onto cross- linked chitosan [31].

In this work, we evaluated the adsorption of chitosan modified with sodium dodecyl sulfate (SDS) as part of the adsorption study of Cr using modified chitosan. SDS-modified chitosan beads have been reported to be effective for removing cationic dyes [32]. Adsorption on surfaces is enabled by the metal ion strength and the presence of key functional groups on the polymer chain [33–35]. The particle aggregation via a bridging structure can be described as a two-step pathway: (1) initial chain adsorption and bridging, followed by (2) floc maturation/ reconfiguration. Before the interparticle connection occurs, the chain of SDS must be adsorbed on a chitosan surface [36]. Furthermore, chitosan modified with SDS has recently been used for the removal of heavy metals, such as cadmium [37, 38]. However, the use of SDS-modified chitosan as a Cr adsorbent, using different initial concentrations of SDS to optimize the adsorbent, has rarely been investigated. The objective of the present research is to determine the efficacy of SDS-modified chitosan beads as a sorbent for Cr(III) and Cr(VI) for future practical applications, as well as to understand the adsorption mechanism. After the characterization of SDS-chitosan by scanning electron microscopy-energy dispersive X-ray spectroscopy (SEM-EDS), X-ray photoelectron spectroscopy (XPS) and Fourier transform-infrared spectroscopy (FT-IR), batch experiments with SDS-modified chitosan beads were carried out to optimize the parameters and to obtain the maximum removal of Cr(III) and Cr(VI). In this study, we examined the effects of several parameters, including solution pH, contact time, adsorbent dosage, and initial concentration.

2. Materials and methods

Chemical reagents, such as chitosan and sodium dodecyl sulfate (SDS; M.W.: 288.372 g/mol), were purchased from Tokyo Chemical Industry Co., Inc. Acetic acid, NaOH, HNO_3, $NaSO_4$, ethylenediaminetetraacetic acid disodium salt dihydrate, and toluene, were purchased from Kanto Chemical Industry Co., Inc. All reagents used were of analytical grade. During the whole working process in this study, the water (>18.2MΩ) treated by the ultrapure water system (RFU 424TA, Advantech Aquarius) was employed. K2CrO7 standard solution (1000 mg·L^{-1} from Kanto Chemical Co., Inc.) was diluted and used to prepare the Cr standard solution for calibration.

2.1 Synthesis of the adsorbent

In this study, chitosan powder with a molecular weight (50–190 kDa) and degree of deacetylation (> 80%) was used. After drying, the viscosity of chitosan was 20 to 100 mPa·s (in 0.5%Acetic acid soln., 20°C). Firstly, 1.5 g of chitosan was placed in acetic acid solution (2.0%), and the solution was mixed for 24 h. The chitosan-gel was prepared by dropping the above chitosan solution into 200 mL of 0.20 mol·L^{-1} NaOH. Consequently, the obtained gel was rinsed with ultrapure water until its pH reached 7 after stirring for 24 h. Secondly, 200 tablets of chitosan-gel beads were placed in 100 mL of SDS solution (including the fixed concentration of SDS), and then left for

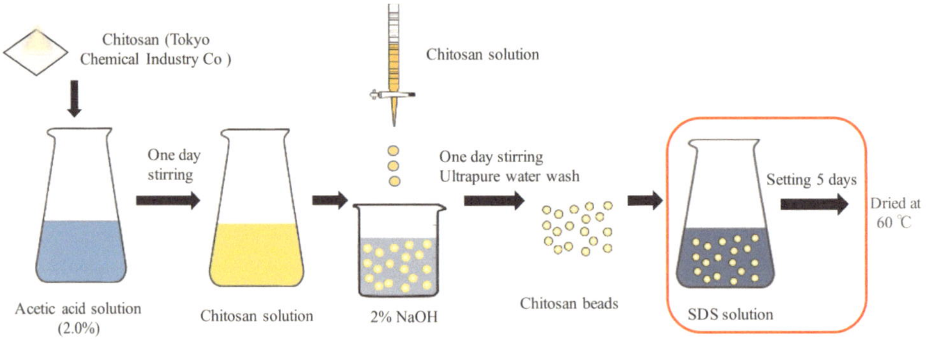

Figure 1.
Procedure employed for the synthesis of sodium dodecyl sulfate (SDS)-chitosan beads.

five days. Thus, chitosan-gel beads modified with SDS were obtained, and finally, they were dried at 60°C overnight for use as an adsorbent. The synthesis procedure for the SDS-chitosan beads is demonstrated in **Figure 1**. It is considered that the prepared adsorbent has a bilayer of SDS over the surface of pure chitosan beads. This bilayer can have a higher ion capturing capacity [37].

2.2 Characterization of the adsorbent

The diameter of chitosan beads was measured to be about 0.5–2 mm (judging from 200 beads as representative chitosan beads). After weighing 200 hydrogel beads, the dry weight per chitosan bead was estimated to be 3.7×10^{-4} g, which suggests that the adsorbent contained 98% moisture. In order to determine the physicochemical properties of pristine and modified chitosan, several characterization methods have been employed. FT-IR spectra of the samples were recorded in the range of 4000–500 cm^{-1} with a JASCO Japan FTIR-4200 spectrophotometer using KBr pellet pressing method. The surface morphology and element distribution of the chitosan beads before and after the adsorption of Cr were observed using SEM-EDS (JEOL Japan: JCM-6000 with JED-2300). X-ray Photoelectron Spectroscopy (XPS, Thermo Scientific Center: K-Alpha) was also used to assess the surface chemistry properties of SDS-modified chitosan beads.

2.3 Adsorption experiments

The beads were put into contact with 50 mL of an aqueous solution containing Cr (III) or Cr (VI) ion with a known initial concentration. The pH of each solution was adjusted using 0.1 mol/L^{-1} NaOH and 0.1 mol/L^{-1} HNO_3. The flask was then placed into an automatic shaker. In the adsorption experiments, the parameters were varied such as the pH range of 1–7, contact time of 1–72 h, SDS initial concentration of 10–9000 mg•L^{-1}, adsorbent dosage of 0.01–0.06 g•L^{-3}, temperature of 288–318 K, and initial Cr(III) or Cr(VI) concentration of 0.1–3.0 mg•L^{-1}. The concentrations of Cr in the filtrate were measured by ICP-MS (Thermo Scientific Center: X-series II). The adsorption capacities of the chitosan beads modified with SDS were calculated by the following Eq. [9]:

$$q_e = (C_i - C_e) \times V/m \tag{1}$$

where q_e represents the adsorption capacity at the equilibrium ($mg{\bullet}g^{-1}$), C_i and C_e are the initial and equilibrium concentrations of Cr in the batch system, respectively ($mg{\bullet}L^{-1}$), V is the volume of the solution (L), and m is the weight of the adsorbents (g).

3. Result and discussion

3.1 Characterization of materials

3.1.1 SEM-EDS micrographs

Figure 2 shows the SEM-EDS images of the chitosan beads and SDS-chitosan. These images showed the unevenness of the surface. This was most likely due to the release of water from the adsorbent during the drying process of chitosan. In addition to the irregularities, SDS-chitosan showed a mesh-like pattern. This was thought to be due to SDS modified on the chitosan surface. In particular, the modified chitosan beads with a high SDS concentration stood out compared to the other beads. As the irregularities on the adsorbent's surface were considered, the adsorption proceeded in

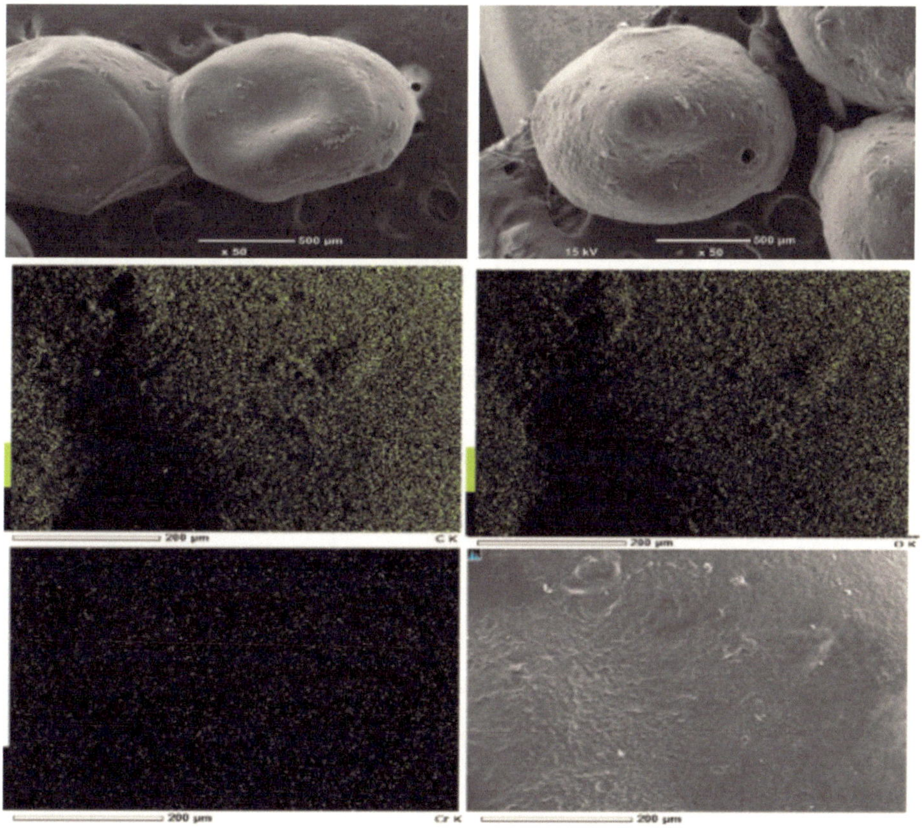

Figure 2.
Scanning electron microscopy (SEM) images of chitosan beads (a) and SDS-chitosan (b), and mapping images of SDS-chitosan after the adsorption of Cr(III).

two ways: physical and chemical adsorption. From the mapping images, it was proven that Cr ions were adsorbed onto the adsorbent surface.

3.1.2 FT-IR and XPS spectra

The surface functional groups and the chemical compositions of the modified SDS-chitosan beads were identified by FTIR and XPS analysis, respectively. The FT-IR results of chitosan and cross-linked chitosan are shown in our previous work [39]. **Figure 3** shows the FT-IR results of SDS-chitosan and the chitosan beads in this study, with the peaks of SDS-chitosan and the chitosan beads can clearly visible. It is apparent from this figure that the main peaks common to each adsorbent were due to the -OH group at 3400–3500 cm^{-1} and the aliphatic methylene group at 2871 cm^{-1}. The amine and ether groups are shown by wide peaks at 1560–1640 cm^{-1} and 1110 cm^{-1}, respectively. For SDS-chitosan, the peak at 1248 cm^{-1} was characteristic of the asymmetrical vibration of the C-O-S group, confirming that the prepared adsorbent was a composite of SDS and chitosan.

The chitosan beads with varying initial loading concentrations of SDS were analyzed using XPS, as shown in **Figure 4**. The C1s spectra of these samples displayed peaks at 284.5, 286.5, and 288.5 eV, corresponding to C-C, C-O, and C=O bonds, respectively. The S2p spectra of SDS600 and 6000-chitosan displayed peaks at 169 eV. The S that seemed to be derived from SDS was also not detected at SDS concentrations of 0 and 100 mg/L but was detected at concentrations of 600 and 6000 mg/L.

3.2 Adsorption experiment

3.2.1 Effect of the initial SDS concentration

To determine the optimum initial loading concentration of SDS for Cr(III) removal, the chitosan beads were modified with SDS solutions ranging from 10 to 9000 mg/L. The adsorption experiments were performed under the following conditions: an initial concentration of Cr(III) of 1 mg/L, a contact time of 2 days, and adsorbent dosage of 0.05 g, a pH of 4, and a temperature of 25°C. In the meantime, initial SDS concentrations were varied from 10 to 6000 mg/L for Cr(VI) removal. The

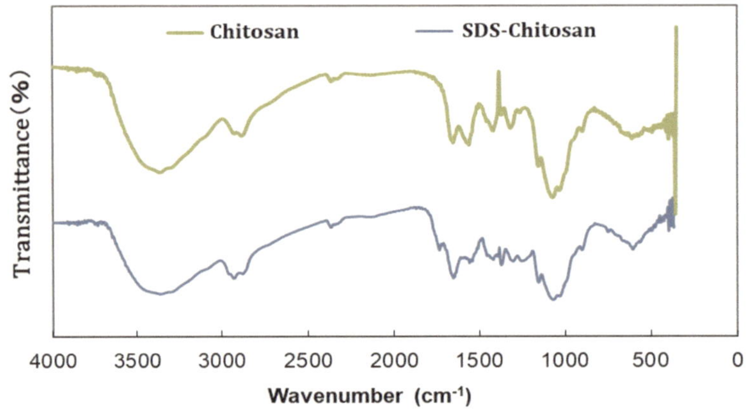

Figure 3.
Fourier transform-infrared spectroscopy (FT-IR) spectra of chitosan and SDS-chitosan.

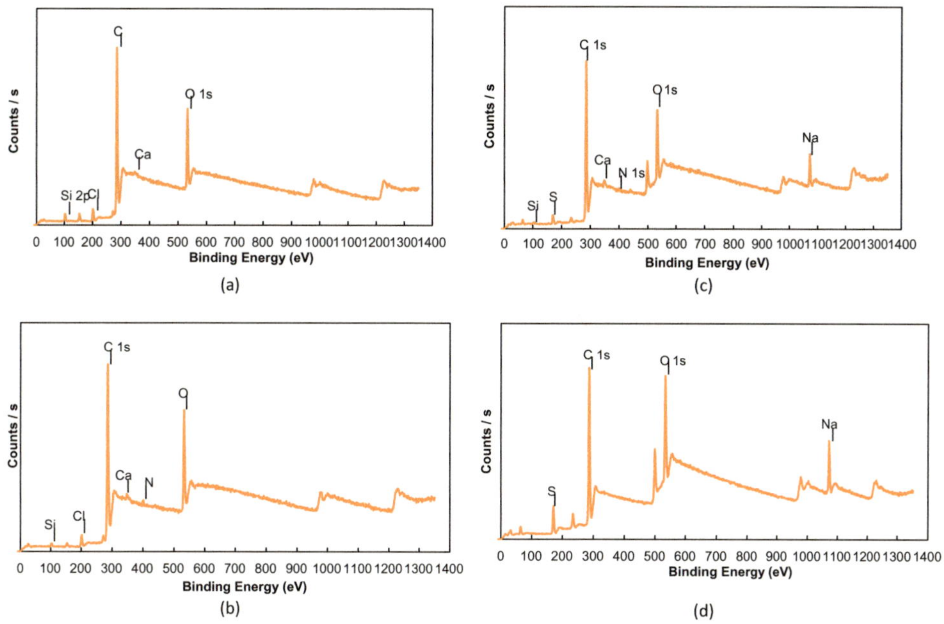

Figure 4.
X-ray photoelectron spectroscopy (XPS) spectra of chitosan (a), SDS100-chitosan (b), SDS600-chitosan (c), and SDS6000-chitosan (d) beads.

experiment was performed under the following conditions: adsorbent dose of 0.05 g, initial concentration of Cr(VI) of 1 mg/L, pH of 4, and contact time of three days.

The results of Cr (III) adsorption are shown in **Figure 5**. The adsorption capacity continuously increased within 6000 mg/L as the SDS concentration increases, but the adsorption amount was almost constant at further higher concentrations. Thus, 6000 mg/L was the optimum initial loading SDS concentration for Cr(III). On the other hand, the adsorption capacity of SDS for the removal of Cr(VI) increased with an increase of the SDS concentration from 10 to 40 mg/L, and after that it decreased as shown in **Figure 6**. The maximum capacity for Cr(VI) was obtained at the initial SDS concentration of 40 mg/L.

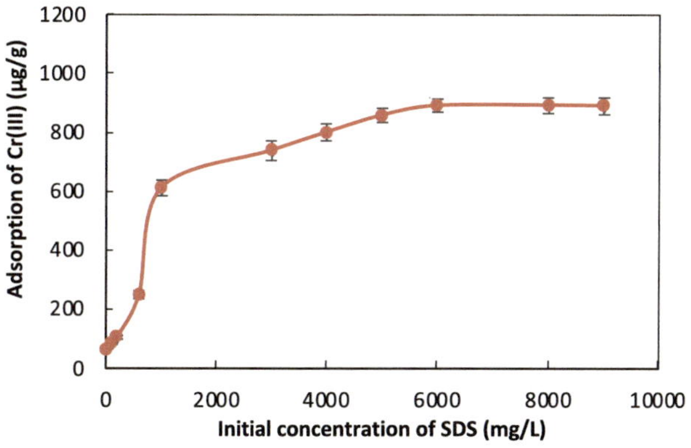

Figure 5.
Effect of the initial SDS concentrations on the adsorption of Cr(III).

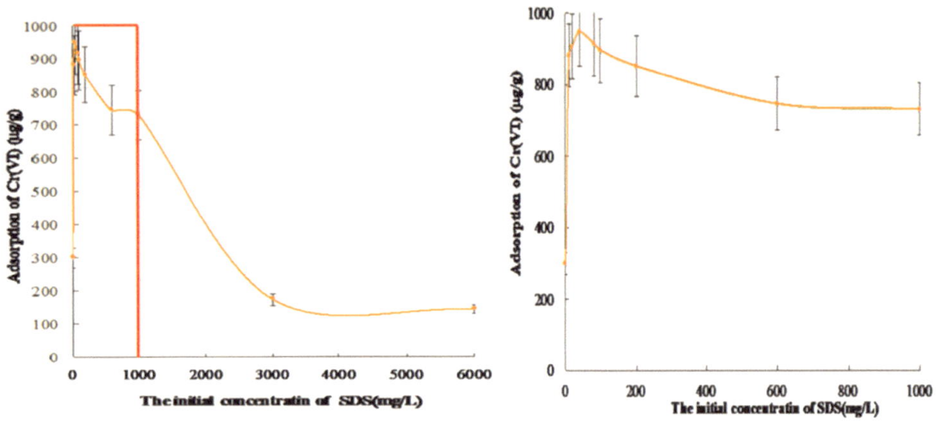

(a) (b)

Figure 6.
Adsorption of Cr(VI) for different initial SDS concentrations (10–6000 mg/L). (a) Effect of initial SDS concentrations (10–6000 mg/L) on the adsorption of chromium (Cr) (VI). (b) Effect of initial SDS concentrations (10–1000 mg/L) on the adsorption of Cr(VI).

3.2.2 Effect of pH

The effect of pH on Cr(III) adsorption by SDS-chitosan was investigated in the pH range of 4–7. Other parameters were set as the following: the contact time was 24 h, the temperature was 25°C, the adsorbent dosage was 0.4 mg/L, and the initial Cr(III) concentration was 1 mg/L. The results are shown in **Figure 7**. As shown in **Figure 7**, the adsorption capacity of Cr(III) increased with the increase of pH from 4 to 7. However, it was found that Cr (III) precipitated as Cr $(OH)_3$ at the pH value in 6 and 7. Thus, a pH of 6 and 7 were not suitable for the adsorption experiments and optimized pH value was 4.

As for the adsorption experiment of Cr(VI), pH was set in the range of 4 to 10 (**Figure 8**). In this experiment, the shaking time was 24 h, the temperature was 25°C, and the dose of adsorbent was 0.02 g/L. As shown in **Figure 8**, pH affects Cr(VI)

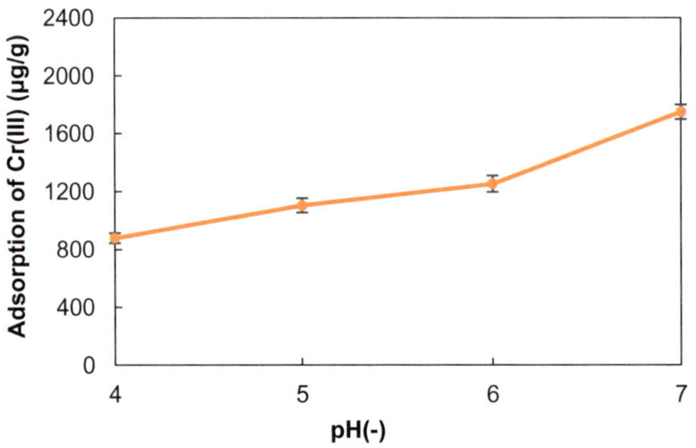

Figure 7.
Effect of pH on the adsorption of Cr(III) by SDS-chitosan.

DOI: http://dx.doi.org/10.5772/intechopen.104093

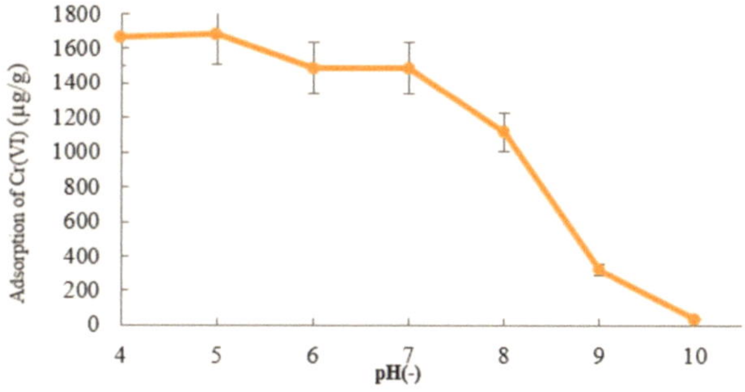

Figure 8.
Effect of pH on the adsorption of Cr(VI) onto the SDS-modified chitosan beads.

adsorption on SDS-modified chitosan beads. The acidity of the solution brought a significant effect on the adsorption of Cr(VI) on SDS-modified chitosan beads where the amino groups of the chitosan were protonated. Cr(VI) ions were most effectively adsorbed at pH 4–5, which may be related to changes in surface charge on the adsorbent. As pH increased above pH 5, the uptake decreased.

3.2.3 Effect of contact time

The influence of contact time on the adsorption of Cr(III) by SDS-chitosan was investigated. The experiment was conducted under the following conditions: a pH of 4, a temperature of 25°C, an adsorbent dosage of 0.05 g, and an initial concentration of Cr(III) of 1 mg/L. **Figure 9** shows that the adsorption capacity of SDS-chitosan for Cr (III) increased sharply within the first 24 h, and continued until the contact time reached 48 h. Thus, 48 h was selected as the optimized contact time.

We have also studied how the contact time affects the adsorption capacities of SDS-modified chitosan beads towards Cr(VI) with varying contact times from 1 to 96 h (**Figure 10**). In this experiment, the concentration of Cr(VI) was set as 1 mg/L with the dose of 0.05 g at a temperature of 25°C adjusted pH to 4.

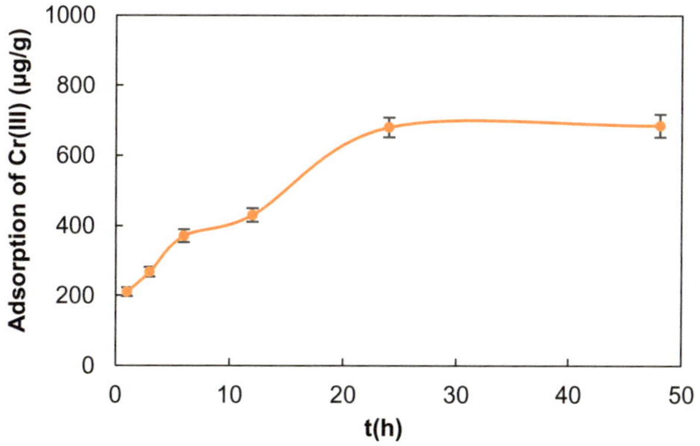

Figure 9.
Effect of contact time on the adsorption of Cr(III) by SDS-chitosan.

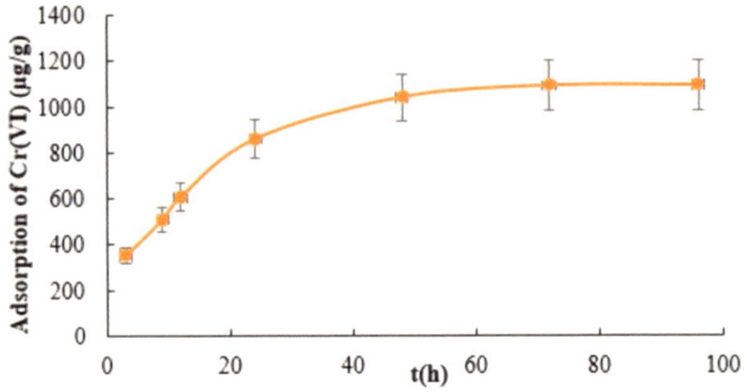

Figure 10.
Effect of the contact time on the adsorption of Cr(VI) on the SDS-modified chitosan beads.

The adsorption capacity of SDS-chitosan beads for Cr(VI) increased sharply within the first 24 h, which may be attributable to the availability of the sites on the surface of the adsorbent. It is suggested that a concentration gradient is present in both the adsorbent and adsorbate in the solution [40]. Then, adsorption reached equilibrium at 72 h, and afterwards, there was no appreciable increase (**Figure 10**). Hence for further studies, the optimized contact time was taken as 72 h.

3.2.4 Effect of the adsorbent dosage

The adsorbent dosage is an important factor that affects the adsorption capacity. To study the effect of adsorbent dosage on the adsorption of Cr(III), adsorption experiments were conducted at a pH of 4, a temperature of 25°C, a contact time of 24 h, and an initial concentration of Cr(III) of 1 mg/L. The results are shown in **Figure 11**. The adsorption rate increased as the adsorbent dosage increased, reaching approximately 70% at 0.8 mg/L, after that, there was no appreciable increase. Thus, 1 mg/L was selected as the optimized adsorbent dosage.

Meanwhile for the adsorption experiment of Cr (VI), 0.8 mg/L was regarded as the optimum dosage. The experiments were performed by varying the dosage (from 0.4 to 1.0 mg/L) and keeping all other parameters constant (temperature: 25°C; pH: 4; contact time: 24 h; initial concentration: 1.0 mg/L). The results are shown in **Figure 12**.

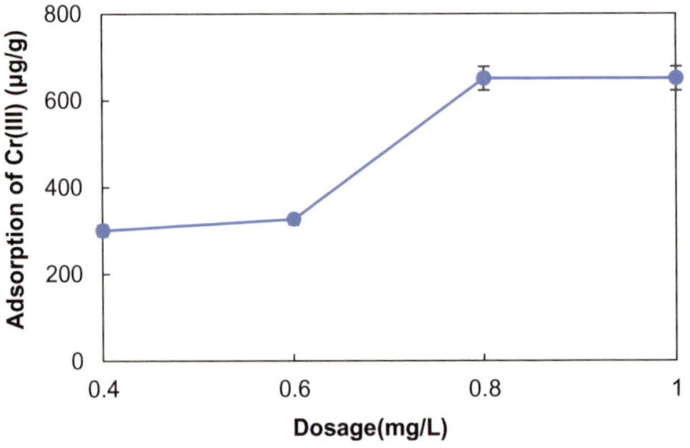

Figure 11.
Effect of the adsorbent dosage on the adsorption of Cr(III) by SDS-chitosan.

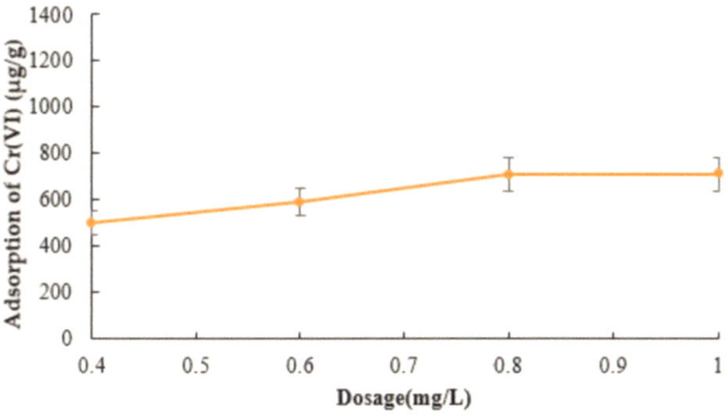

Figure 12.
Effect of the adsorbent dosage on the adsorption of Cr(VI) on the SDS-modified chitosan beads.

3.2.5 Effect of competitive ions

In this study, adsorption experiments of Cr(III) were conducted in the presence of several competitive ions with different concentrations (0, 50, 100, and 200 mg/L) of Na^+, Ca^{2+}, Ba^{2+}, K^+, Mg^{2+}, and Sr^{2+}. The initial concentration of Cr(III) was set to 1.0 mg/L, the pH was 4, the temperature was 25°C, the contact time was 24 h, and the adsorbent dosage was 1.0 mg/L. The effect of competitive ions on the adsorption of Cr (III) is shown in **Figure 13**. It was confirmed that the adsorption capacity of Cr(III) did not decrease at all even in the presence of other metal ions.

Conversely, it was found that Cr(VI) adsorption was inhibited when the concentrations of coexisting ions were high. The effect of competitive anions on the adsorption of Cr(VI) is shown in **Figure 14**. In this study, the initial concentration of Cr(VI) was fixed to 1 $mg{\cdot}L^{-1}$. These counter ions were tested collectively, and all the ions were included at 50, 100, or 200 mg/L in solution. From this figure, the removal of Cr (VI) was remarkably reduced under the presence of common ions at above 50 $mg{\cdot}L^{-1}$ (i.e., 50 times the Cr(VI) concentration or more), although no substantial decrease was observed when the concentration of each common ion was less than 10 $mg{\cdot}L^{-1}$ in our previous preliminary experiments.

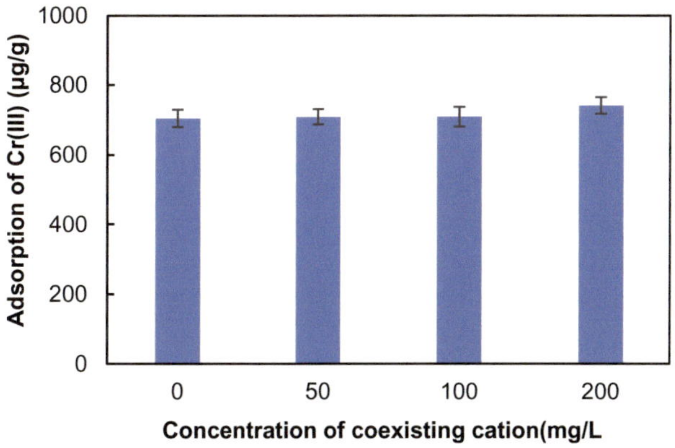

Figure 13.
The effect of competitive ions on the adsorption of Cr(III).

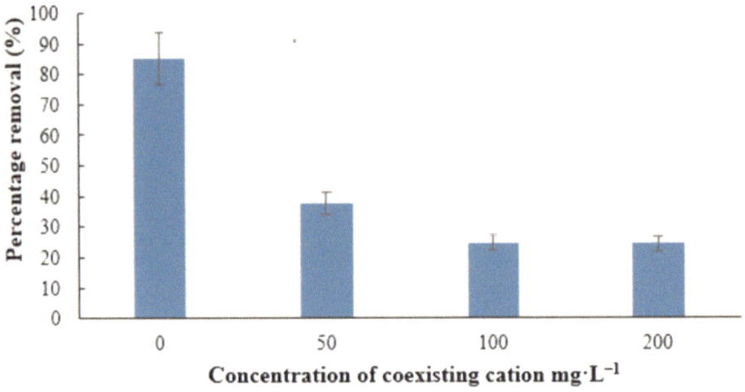

Figure 14.
Effect of competitive anions on the adsorption of Cr(VI) on the SDS-modified chitosan beads.

3.3 Adsorption isotherms

Adsorption isotherms describe the interactive process between the adsorbents and adsorbates in aqueous medium at the attained saturation point. Adsorption isotherms of Cr(III) and Cr(VI) on SDS-chitosan were identified with different initial concentrations from 0.01 to 2 mg/L under optimized conditions for the pH (4), contact time (48 h for Cr(III), 72 h for Cr(VI)), and dosage of the adsorbent (1 mg/L for Cr(III), 0.8 mg/L for Cr(VI)). The adsorption of Cr on SDS-chitosan was evaluated using typical adsorption isotherms, the Langmuir and Freundlich models (**Figure 15(a)** and **(b)**).

The adsorption data obtained for Cr(III) using SDS-chitosan were analyzed by Langmuir and Freundlich equations (**Figure 16(a)** and **(b)**).

Langmuir equation:

$$\frac{C_e}{q_e} = \frac{C_e}{q_{max}} + \frac{1}{K_L q_{max}} \tag{2}$$

Freundlich equation:

$$lgq_e = lgK_F + (1/n)lgC_e \tag{3}$$

where C_e and q_e are the concentration of Cr at the equilibrium (mg•L^{-1}) and the amount of adsorption of Cr(III) at the equilibrium (mg•g^{-1}), respectively, q_{max} is the maximum adsorption capacity on the surface of the chitosan bead (mg•g^{-1}), K_L is the Langmuir constant related to the adsorption strength or intensity (L•mg^{-1}), K_F is the Freundlich constant and $1/n$ indicates the adsorption intensity of the system [41, 42].

The Langmuir isotherm model assumes that a monolayer adsorption occurs on the surface of an adsorbent. The slope of the linearized Langmuir isotherm can be used to explain the type of sorption using the Hall separation factor (R_L), which is favorable ($0 < R_L < 1$), unfavorable ($R_L < 0$), linear ($R_L = 1$) or irreversible ($R_L = 0$) [40]. The Freundlich isotherm model assumes a multilayer adsorption process without considering the adsorption saturation, which occurs on the multilayer heterogeneous surface. The value of the adsorption intensity of the system($1/n$) can be used to judge the difficulty of the adsorption process: irreversible ($1/n = 0$), favorable ($0 < 1/n < 1$), or unfavorable ($1/n > 1$) [43]. **Table 1** displays the correlation coefficient (R^2) of these isotherms for Cr(III) on SDS-chitosan along with other relevant parameters.

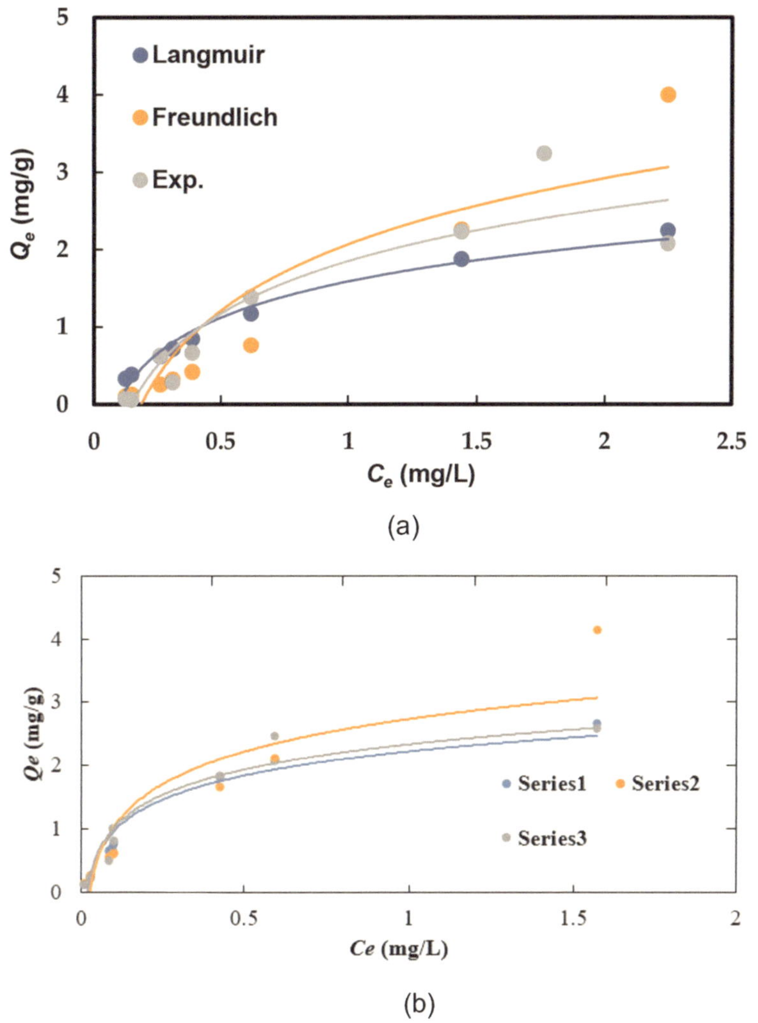

Figure 15.
***(a)**. Adsorption isotherms of Cr(III) using SDS-chitosan. **(b)**. Adsorption isotherms of Cr(VI) using SDS-chitosan.*

Metal T(K)		Langmuir Isotherm			Freundlich Isotherm		
		Q_{max} (mg/g)	R_L	R^2	K_F (mg/g)	*1/n*	R^2
Cr(III)	298	3.42	1.15×10^{-3}	0.842	1.41	1.21	0.820
Cr(VI)	298	3.23	0.31×10^{-4}	0.960	3.01	0.700	0.921

Table 1.
Isotherm parameters for Cr(III) and Cr(VI) adsorption onto SDS-chitosan.

As shown in **Figure 16(a)**, **(b)**, and **Table 1**, it was revealed that the R^2 values for Cr(III) and Cr(VI) were relatively large and a favorable adsorption of Cr(III) and Cr (VI) on SDS-chitosan was presented. Compared to the Freundlich isotherm, the Langmuir isotherm had a larger R^2value, which suggests that the adsorption of Cr(III)

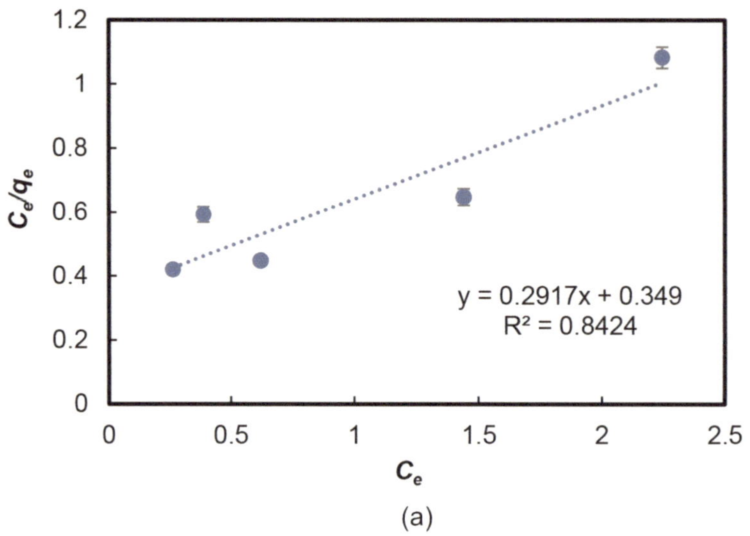

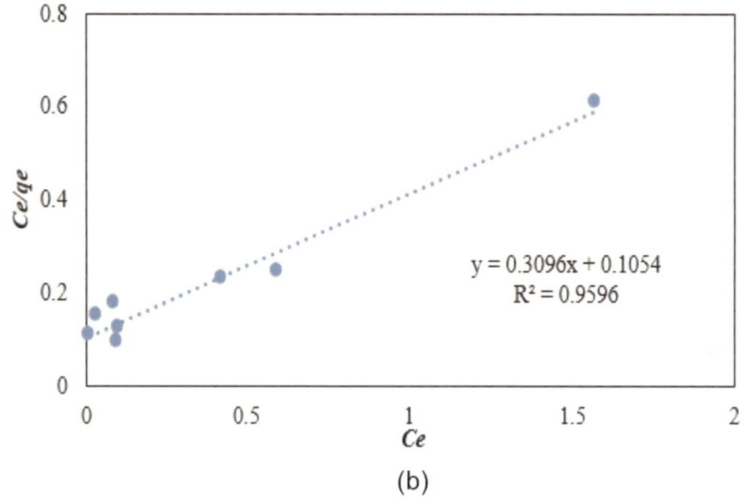

Figure 16.
***(a).** Langmuir isotherm of Cr(III) adsorption onto SDS-chitosan. **(b).** Langmuir isotherm of Cr(VI) adsorption onto SDS-chitosan.*

and Cr(VI) on SDS-chitosan mainly occurred by a monolayer reaction. The maximum adsorption capacity (q_{max}) calculated from the Langmuir model was 3.42 mg/g for Cr (III) and 3.23 mg/g for Cr(VI).

3.4 Kinetic studies

The rate-controlling steps of the adsorption system are essential to survey the mechanism of Cr(III) and Cr(VI). Kinetic studies were conducted to explain the adsorption mechanism of Cr(III) and Cr(VI) ions onto the SDS-chitosan beads. The effects of contact time on the kinetics of Cr(III) and Cr(VI) adsorption by SDS-chitosan adsorbent are displayed in **Figure 17(a)** and **(b)**. According to **Figure 17(a)**, the removal of Cr(III) from SDS-chitosan increased sharply in the initial 6 h, indicating that the uptake of Cr(III) was primarily caused by chemical sorption. As shown in

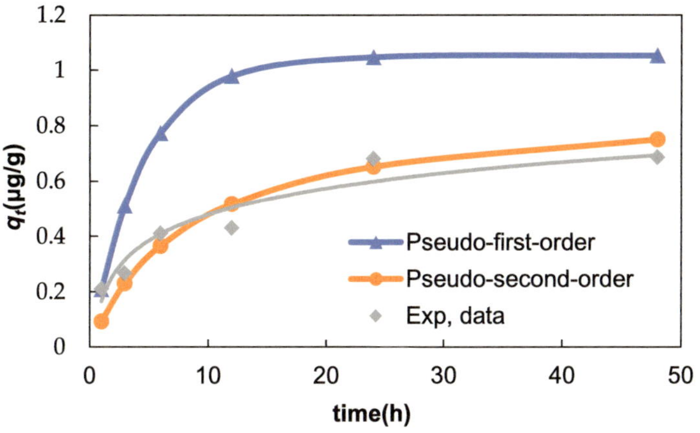

(a)

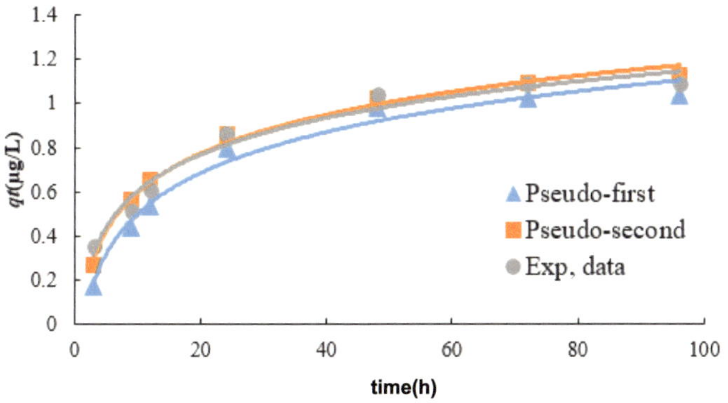

(b)

Figure 17.
(a). *Pseudo-first-order and pseudo-second-order Cr(III) adsorption on SDS-chitosan.* ***(b).*** *Pseudo-first-order and pseudo-second-order Cr(VI) adsorption on SDS-chitosan.*

Figure 17(b), the adsorption of Cr(VI) by SDS-chitosan beads increased significantly within 96 hours, although the increase continued at a slower rate for 48 h. The rapid adsorption within the initial 24 h indicated that Cr(VI) uptake was mainly dominated by chemical sorption or surface complexation.

To investigate the mechanism of adsorption of Cr(III) and Cr(VI) on SDS-chitosan, fitting was determined according to the pseudo-first-order (**Figure 18(a)** and **(b)**), pseudo-second-order (**Figure 19(a)** and **(b)**) and intraparticle diffusion kinetic models (**Figure 20(a)** and **(b)**). The equations of these models are given by:

Pseudo-first-order model:

$$ln = ln\left(q_e\right) - k_1 t \tag{4}$$

Pseudo-second-order model:

$$\frac{t}{q_t} = \frac{1}{k_2 q_e^2} + \frac{t}{q_e} \tag{5}$$

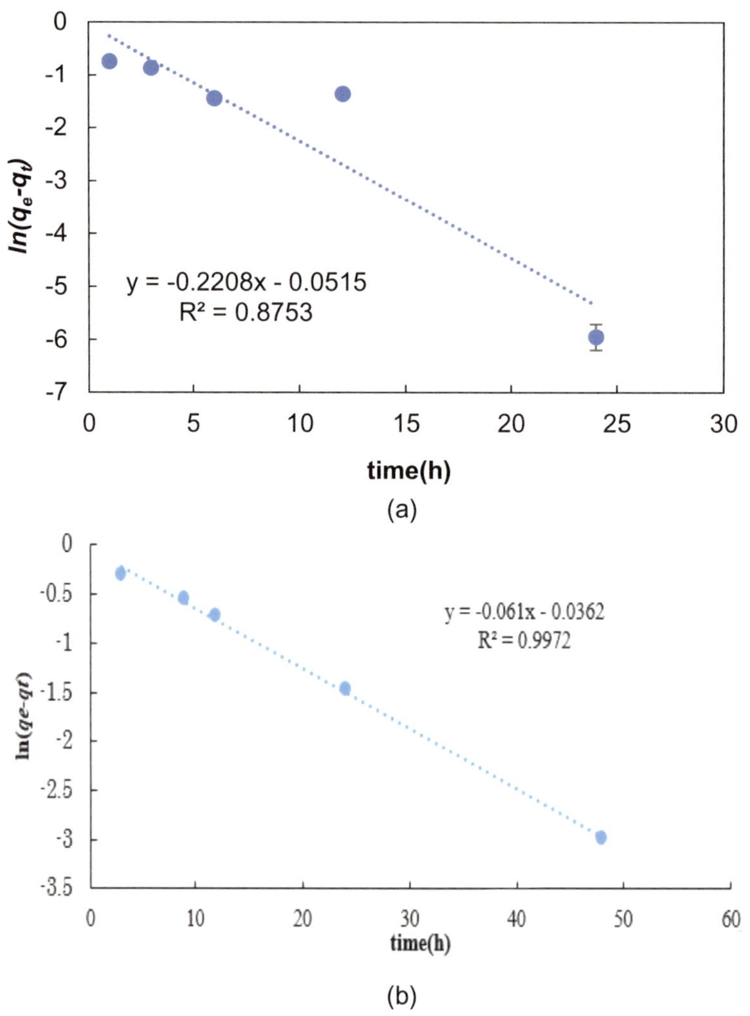

Figure 18.
(a). *Pseudo-first-order linear kinetic model of Cr(III) adsorption by SDS-chitosan.* ***(b).*** *Pseudo-first-order linear kinetic model of Cr(VI) adsorption by SDS-chitosan.*

Intraparticle diffusion model:

$$q_t = K_{id}t^{1/2} \tag{6}$$

where q_e and q_t are the adsorption capacities of Cr using the SDS-chitosan beads at the equilibrium and time, t, respectively (mg•g^{-1}), $k_1$is the rate constant of the pseudo-first-order adsorption (h^{-1}), k_2 is the rate constant of the pseudo-second-order adsorption (g•mg^{-1}•h^{-1}) and K_{id} is the rate constant of intraparticle diffusion (mg•g^{-1}•h$^{-1/2}$) [44–46].

The sorption kinetic parameters including k_1, k_2, K_{id}, q_e and the correlation coefficients R^2 are presented in **Table 2**. From **Table 2**, the correlation coefficient R^2 for the pseudo-second-order model was higher than that for the pseudo-first-order and intraparticle diffusion models. Based on these results, the sorption kinetics of

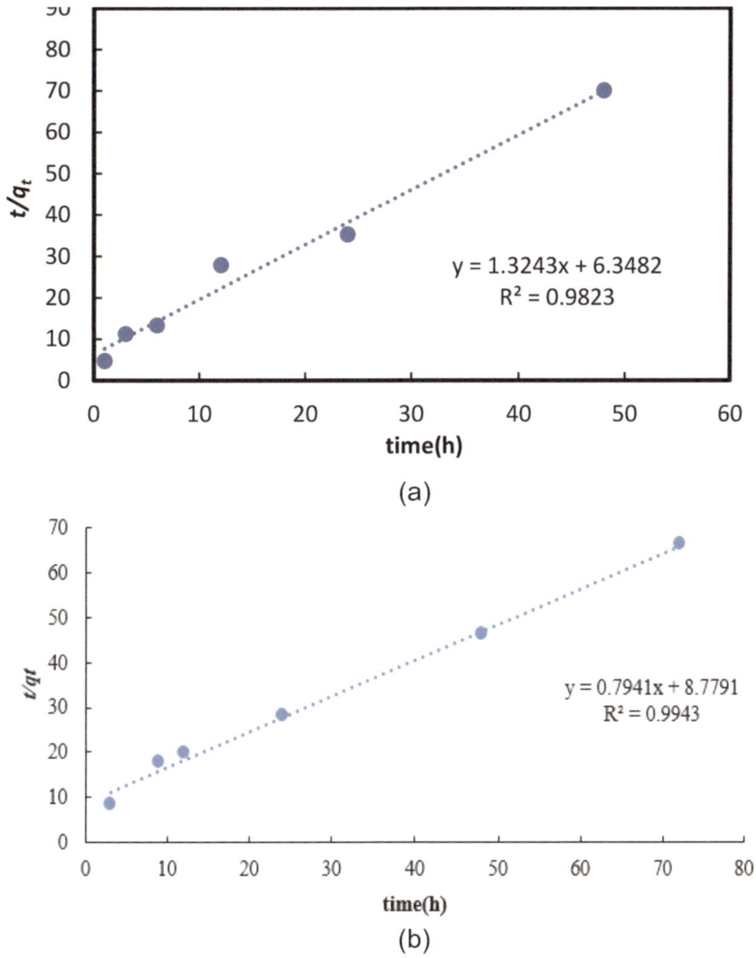

Figure 19.
***(a)**. Pseudo-second-order linear kinetic model of Cr(III) adsorption by SDS-chitosan. **(b)**. Pseudo-second-order linear kinetic model of Cr(VI)adsorption by SDS-chitosan.*

Metal	q_e (mg/g)	Pseudo-first-order model			Pseudo-second-order model			Intraparticle Diffusion	
		q_e (mg/g)	k_1 (h^{-1})	R^2	q_e (mg/g)	k_2 ($g{\cdot}mg^{-1}{\cdot}h^{-1}$)	R^2	K_{id}	R^2
Cr(III)	0.69	1.05	0.22	0.88	1.26	0.07	0.98	0.17	0.90
Cr(VI)	1.09	1.04	0.06	0.99	0.88	0.13	0.99	0.09	0.99

Table 2.
Kinetic parameters of Cr(III) and Cr(VI) absorption onto SDS-chitosan.

the Cr(III) ion on SDS-chitosan can be described by the pseudo-second-order model, indicating that the chemical adsorption may be the rate-limiting step of the process. Assuming adsorbent particle is sphere of radius ‘a’, the diffusion follows Fick’s law that was applied to confirm it. In the law, the relationship between adsorption capacity and time given by the following Equations [44].

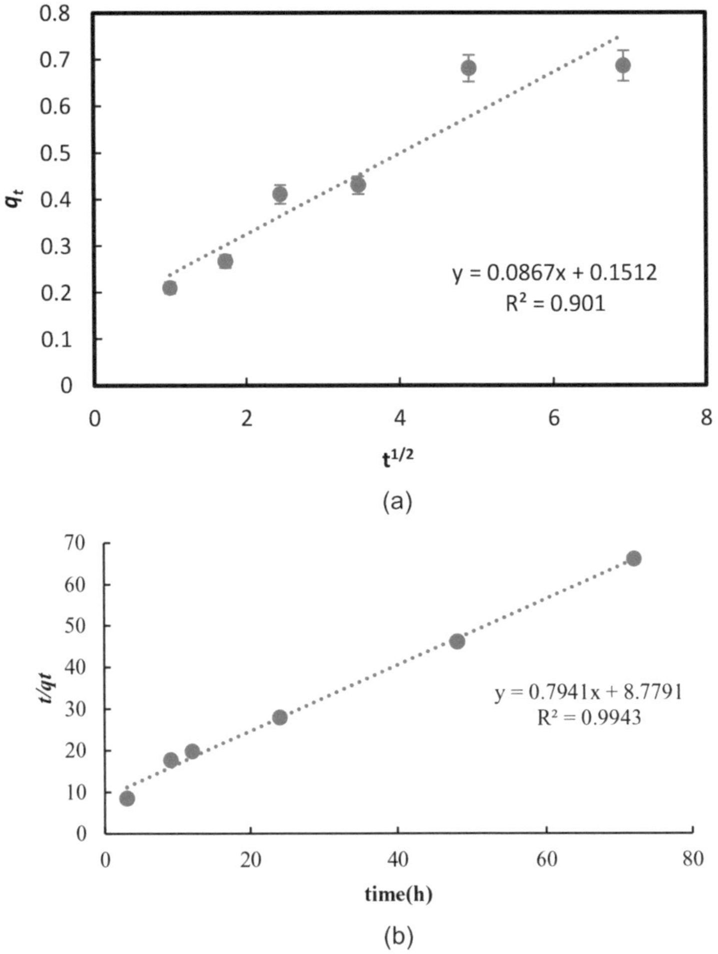

Figure 20.
***(a)**. Intraparticle diffusion kinetic model of Cr(III) absorption by SDS-chitosan. **(b)**. Intraparticle diffusion kinetic model of Cr(VI) absorption by SDS-chitosan.*

$$\frac{q_t}{q_e} = 6\left(\frac{D_t}{a^2}\right)^{1/2}\left\{\pi^{-1/2} + 2\sum_{n=1}^{\infty} ierfc \frac{na}{\sqrt{(Dt)}}\right\} - 3\frac{Dt}{a^2} \tag{7}$$

After D is replaced with D_1, at a short time, Eq. (7) becomes as:

$$\frac{q_t}{q_e} = 6\left(\frac{D_1}{\pi a^2}\right)^{1/2} t^{1/2} \tag{8}$$

The film diffusion coefficient D_1 is calculated from the slope of the plot of q_t/q_e versus $t^{1/2}$. For moderate and large times, the diffusion equation given by

$$\frac{q_t}{q_e} = 1 - \frac{6}{\pi^2}\sum_{n=1}^{\infty}\frac{1}{n^2} exp\left(\frac{-Dn^2\pi^2 t}{a^2}\right) \tag{9}$$

C_0 (mg/L)	$D_1 \cdot 10^{-8}$ (cm^2 /s)	$D_2 \cdot 10^{-12}$ (cm^2 /s)
1.00	7.04	1.51

Table 3.
The film and pore diffusion coefficients of the adsorption.

If D is replaced with D_2, at the large time the Eq. (9) becomes as:

$$ln\left(1-\frac{q_t}{q_e}\right)=ln\,\frac{6}{\pi^2}-\left(\frac{D_2\pi^2}{a^2}t\right) \tag{10}$$

The pore diffusion coefficient D_2 is obtained from the slope of the plot of $ln(1-q_t/q_e)$ versus t. The values of D_1 and D_2 are presented in **Table 3**.

It is also known that if the adsorption is controlled by the film diffusion, the values of the film diffusion coefficient may be in the range of $10^{-6}-10^{-8}$ cm^2 /s. While if the adsorption mechanism is controlled by the pore diffusion mechanism, the value of the pore diffusion coefficient belongs to the range of 10^{-11}–10^{-13} cm^2/s [45]. As can be seen in **Table 3**, The D_1 and D_2 values in the model based on the Fick's law were bigger than that above mentioned ranges. It is confirmed that the sorption kinetics of the Cr(III) ion on SDS-chitosan could be described by the pseudo-second-order model.

3.5 Thermodynamic study

To explore the effect of temperature on Cr(III) adsorption by SDS-chitosan, adsorption experiments were performed at temperatures ranging from 298 to 318 K. The results are displayed in **Figures 21** and **22**. In these temperature ranges, the amount of Cr(III) adsorbed on SDS-chitosan increased with the increase of temperature.

Furthermore, based on the experimental results, the thermodynamic parameters of the adsorption, such as standard Gibb's free energy change (ΔG^0), a change in

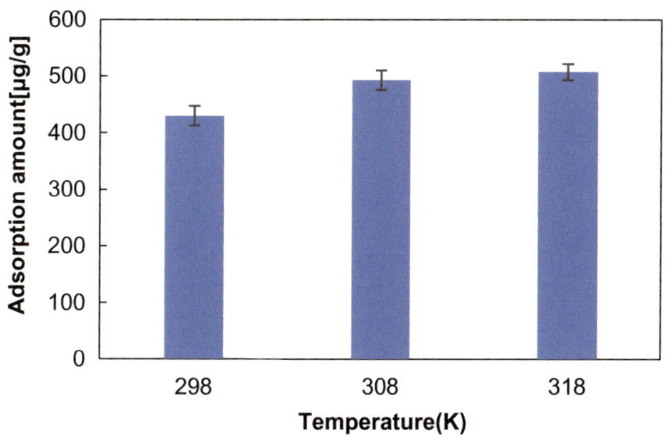

Figure 21.
Effect of temperature on the adsorption of Cr(III) onto SDS-chitosan.

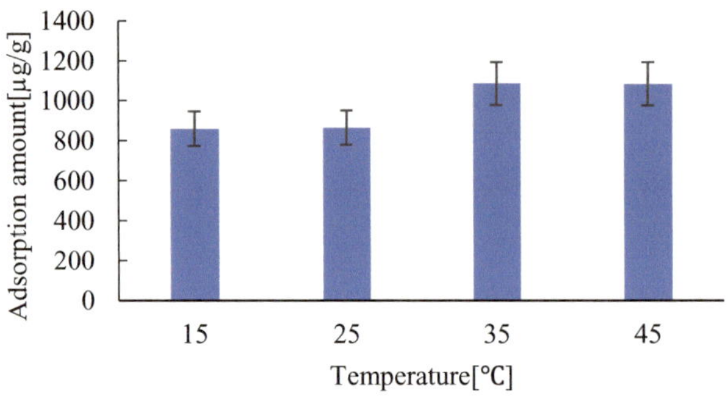

Figure 22.
Effect of temperature on the adsorption of Cr(VI) on the SDS-modified chitosan beads.

standard enthalpy (ΔH^0) and standard entropy (ΔS^0), were determined from Van 't Hoff Equation [46–48].

$$\Delta G^0 = -RTlnK_c \tag{11}$$

$$ln\,K_c = \frac{\Delta H^0}{-RT} + \frac{\Delta S^0}{R} \tag{12}$$

$$\Delta G^0 = \Delta H^0 - T\Delta S^0 \tag{13}$$

where R is the universal gas constant (8.314 J/mol^{-1}•K^{-1}) and T is the temperature (K). The slope and intercept of the plot of $ln\,K_c$ versus $1/T$ were used to determine the ΔH^0 and ΔS^0.

Tables 4 and **5** indicate the thermodynamic parameters for Cr(III) and Cr(VI), respectively. In the temperature range of 298–318 K, the values of $\Delta G > 0$ were obtained. The values of ΔG at various temperatures indicated the feasibility of the process. This suggested that the adsorption process was more spontaneous at higher

T (K)	ΔH (kJ/mol)	ΔS (J/mol)	ΔG (kJ/mol)
298	—	—	0.855
308	12.0	37.4	0.481
318	—	—	0.107

Table 4.
Thermodynamic parameters for the adsorption of Cr(III) on SDS-chitosan.

T(K)	ΔH(kJ/mol)	ΔS(J/mol)	ΔG(kJ/mol)
288	80.70	288.18	−2.34
298	—	—	−5.22
308	—	—	−8.10
318	—	—	−10.98

Table 5.
Thermodynamic parameters for the adsorption of Cr(VI) on SDS-modified chitosan beads.

temperatures. Positive Δ*H* and Δ*S* values also suggest that the sorption process is endothermic, and that the disorderliness of the sorption may occur at the solid-liquid interface. It was then proposed that the adsorption of Cr(III) on SDS-chitosan could be mainly dominated by chemisorption.

3.6 Desorption study

Repeated use of adsorbent and recovery of the adsorbed metal ions are important indicators for evaluating economic efficiency. In this study, regeneration experiments were carried out using the SDS-chitosan beads after adsorption of Cr(III) and Cr(VI). After adsorption, the spent adsorbent was treated with 50 ml of 0.1 mol/L NaOH solution and ultrapure water as desorption agent in desorption experiment, and then filtered. The content of Cr(III) and Cr(VI) in the filtrate was determined by ICP-MS [49]. **Figure 23**, Cr(III) by using NaOH or pure water, respectively. As can be seen in **Figure 23**, the desorption efficiency of Cr(III) was calculated as 57% when using 0.1 mol/L NaOH. The desorption of NaOH was larger than ultrapure water. This indicated that NaOH could be a desorption agent for Cr(III). As shows in **Figure 24**, the desorption efficiency of Cr(VI) was found to be 50%

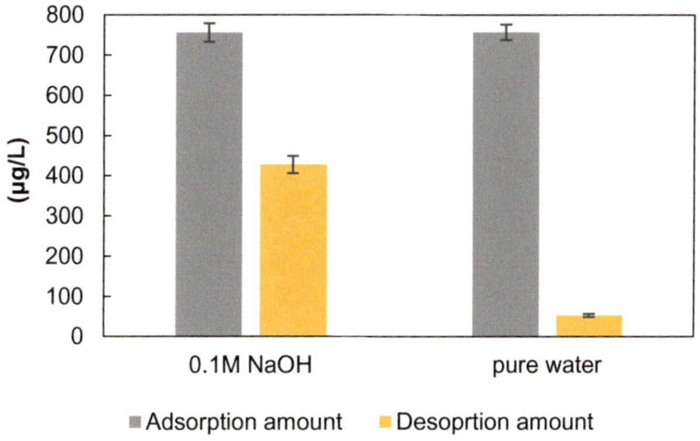

Figure 23.
Desorption efficiency of Cr(III) using NaOH or pure water.

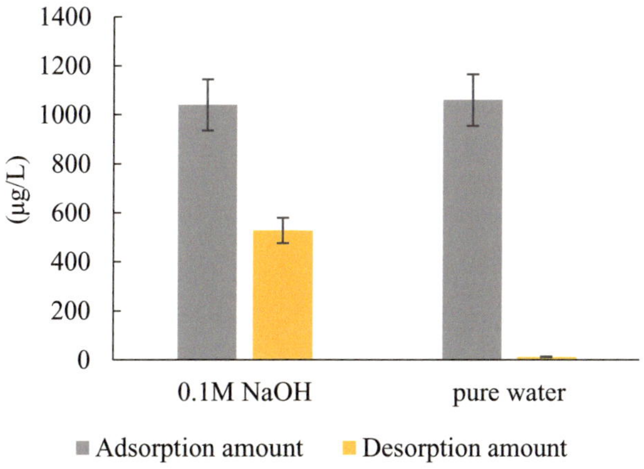

Figure 24.
Desorption efficiency of Cr(VI) using NaOH or pure water.

when 0.1 mol/L NaOH was used, and the desorption was considerably lower with H_2O than NaOH. It is suggested that NaOH can be used as a desorption agent for Cr(VI), although a further investigation is needed for the effective recovery and recycling of Cr(VI).

3.7 Comparison with other adsorbents

The comparison of the maximum adsorption capacity of Cr(III) by SDS-chitosan in the present study with those of other adsorbents in the literature is presented in **Table 6**. As shown in **Table 6**, the adsorption capacity of SDS-chitosan for Cr(III) in this work was not necessarily high compared to other adsorbents. Meanwhile, **Table 7** summarizes the comparison of maximum adsorption capacity of Cr(VI) on SDS-chitosan beads with other adsorbents. It can also be noted that the adsorption capacity is not so high compared to other adsorbents. However, it could be regarded as a potential adsorbent for treating Cr(III) and Cr(VI) from wastewater for practical usage because the synthesis method of the adsorbent is relatively simple.

Adsorbent	Adsorption capacity ($mg \cdot g^{-1}$)	References
Vesicular basalt	0.98	[50]
Chitosan	1.67	[51]
Attapulgite	11.0	[51]
Chitosan/attapulgite	27.0	[51]
Crystalline hydrous titanium(IV) oxide	14.0	[52]
Natural sepiolite	0.53	[53]
Borax sludge	4.00	[54]
SDS-chitosan	3.42	This study

Table 6.
Comparison of adsorption capacity for Cr(III) by different adsorbents.

Adsorbents	Adsorption Capacity ($mg \cdot g^{-1}$)	Reference
Magnetic Chitosan	69.4	[14]
Carboxymethyl Chitosan-Silicon Dioxide	80.7	[30]
Chitosan-g-poly/silica	55.7×10^{-3}	[55]
Crosslinked chitosan bentonite composite	89.1	[56]
Graphene oxide/chitosan	86.2	[57]
Ethylenediamine-magnetic chitosan	51.8	[55]
SDS-chitosan	3.23	This study

Table 7.
The comparison of adsorption properties of several adsorbents.

4. Conclusions

In this study, sodium dodecyl sulfate (SDS) was used to chemically modify chitosan to enhance its adsorption capacity for the removal of chromium. SEM-EDS, FT-IR, and XPS were used to characterize the SDS-chitosan beads. The effect of important operating parameters, such as the loading amounts of SDS, solution pH, contact time, adsorbent dose, temperature, and initial Cr(III) or Cr(VI) concentration, on the adsorption performance was examined in a batch system. The experimental data were found to be fit best using Langmuir isotherm and pseudo-first-order kinetic models. At pH 4–5 and higher temperatures, the adsorption process performed admirably. The maximum adsorption capacity of Cr(III) and Cr(VI) on SDS modified chitosan beads were estimated to 3.42 $mg{\cdot}g^{-1}$ and 3.23 $mg{\cdot}g^{-1}$, respectively in this work. Finally, the SDS-chitosan beads synthesized in this work can be effectively utilized to remove chromium successfully.

Acknowledgements

The authors are grateful to Mr. H. Morohashi of the Industrial Research Institute of Niigata Prefecture for the measurement of XPS and useful advice. The authors also thank Mr. M. Ohizumi of the Office for Environment and Safety, and Mr. N. Miyamoto, Dr. M. Teraguchi, Mr. T. Nomoto, and Prof. T. Tanaka, of the Facility of Engineering in Niigata University for permitting the use of ICP-AES, ICP-MS, FT-IR, and SEM-EDS.

Funding

The present work was partially supported by a Grant-in-Aid for Scientific Research from the Japan Society for the Promotion of Science (Research Program (C), No. 21 K12290).

Conflicts of interest

The authors declare no conflict of interest.

Author details

Naoki Kano[1*], Zou Ming[2], David Eva Vanessa Anak[2] and Muhammad Nabil Md Sari[2]

1 Department of Chemistry and Chemical Engineering, Niigata University, Niigata, Japan

2 Graduate School of Science and Technology, Niigata University, Niigata, Japan

*Address all correspondence to: kano@eng.niigata-u.ac.jp

References

[1] Madala S, Nadavala SK, Vudagandla S, Boddu VM, Abburi K. Equilibrium, kinetics, and thermodynamics of cadmium (II) biosorption on to composite chitosan biosorbent. Arabian Journal of Chemistry. 2017;**10**:1883-1893

[2] Han SL, Zang YA, Gao Y, Yue QY, Zhang P, Kong WJ, et al. Co-monomer polymer anion exchange resin for removing Cr (VI) contaminants: Adsorption kinetics, mechanism and performance. Science Total Environment. 2020;**709**:1-10

[3] Pietrelli L, Francolini I, Piozzi A, Sighicelli M, Silvestro I, Vocciante M. Chromium (III) removal from wastewater by chitosan flakes. Applied Sciences. 1925;**2020**:10

[4] Pietrelli L, Ippolito NM, Reverberi AP, Vocciante M. Heavy metals removal and recovery from hazardous leather sludge. Chemical Engineering Transactions. 2019;**76**: 1327-1332

[5] Rajapaksha AU, Alam MS, Chen N, Alessi DS, Igalavithana AD, Tsang DCW, et al. Removal of hexavalent chromium in aqueous solutions using biochar: Chemical and spectroscopic investigations. Science Total Environment. 2018;**625**:1567-1573

[6] Shi S, Yang J, Liang S, Li M, Gan Q, Xiao K, et al. Enhanced Cr (VI) removal from acidic solutions using biochar modified by Fe3O4 @ SiO2-NH2 particles. Science Total Environment. 2018;**628**:499-508

[7] Wang Y, Zhang N, Chen D, Ma D, Liu G, Zou X, et al. Facile synthesis of acid-modified UiO-66 to enhance the removal of Cr (VI) from aqueous solutions. Science Total Environment 2019, 682, 118–127

[8] Fendorf SE. Surface reactions of chromium in soils and waters. Geoderma. 1995;**67**:55-71

[9] Yamada K, Ishiguro Y, Kimura Y, Asamoto H, Minamisawa H. Two-step grafting of 2-hydroxyethyl methacrylate (HEMA) and 2-(dimethylamino)ethyl methacrylate (DMAEMA) onto a polyethylene plate for enhancement of Cr (VI) ion adsorption. Environmental Technology. 2019;**40**:855-869

[10] Chour Z, Laubie B, Morel JL, Tang Y, Qiu RL, Simonnot MO, et al. Recovery of rare earth elements from Dicranopteris dichotoma by an enhanced ion exchange leaching process. Chemical Engineering and Processing. 2018;**130**: 208-213

[11] Fu F, Wang Q. Removal of heavy metal ions from wastewater: A review. Journal of Environmental Management. 2011;**92**:407-418

[12] Rahmati MM, Rabbani P, Abdolali A, Keshtkar AR. Kinetics and equilibrium studies on biosorption of cadmium, Lead and nickel ions from aqueous solutions by intact and chemically modified Brown algae. Journal of Hazardous A Materials. 2011;**185**:401-407

[13] Yang MJ, Liang XL, Ma LY, Huang J, He HP, Zhu JX. Adsorption of REEs on kaolinite and halloysite: A link to the REE distribution on clays in the weathering crust of granite. Chemical Geology. 2019;**525**:210-217

[14] Huang G, Zhang H, Shi JX, Langrish TSG. Adsorption of chromium (VI) from aqueous solutions using

cross-linked magnetic chitosan beads. Journal of Industrial & Engineering Chemical Research. 2009;**48**:2646-2651

[15] Ngah WS, Ghani SA, Kamari A. Adsorption behaviour of Fe (II) and Fe (III) ions in aqueous solution on chitosan and cross-linked chitosan beads. Bioresource Technology. 2005;**96**: 443-450

[16] Wang G, Liu J, Wang X, Xie ZY, Deng N. Adsorption of uranium (VI) from aqueous solution onto cross-linked chitosan. Journal of Hazardous Materials. 2009;**168**:1053-1058

[17] Inoue K, Fingerman M, Nagabhushanam R, Thompson M. Application of chitosan in separation and purification of metals. Recent Advances Marine Biotechnology Environment Marine Biotechnology. 1998;**2**:63-97

[18] Guibal E, Larkin A, Vincent T, Tobin JM. Chitosan sorbents for platinum sorption from dilute solutions. Industrial and Engineering Chemistry Research. 1999;**38**:4011-4022

[19] Guzman J, Saucedo I, Revilla J, Navarro R, Guibal E. Vanadium interactions with chitosan: Influence of polymer protonation and metal speciation. Langmuir. 2002;**18**:1567-1573

[20] Ng JCY, Cheung WH, McKay G. Equilibrium studies of the sorption of Cu (II) ions onto chitosan. Journal of Colloid and Interface Science. 2002;**255**:64-74

[21] Arrascue ML, Garcia HM, Horna O, Guibal E. Gold sorption on chitosan derivatives. Hydrometallurgy. 2003;**71**: 191-200

[22] Gerente C, Lee VC, Cloirec PL, McKay G. Application of chitosan for the removal of metals from wastewaters by adsorption-mechanisms and models review. Critical Reviews in Environmental Science and Technology. 2007;**37**:41-127

[23] Repoa E, Warchoł JK, Bhatnagar A, Sillanpää M. Heavy metals adsorption by novel EDTA-modified chitosan–silica hybrid materials. Journal of Colloid and Interface Science. 2011;**358**:261-267

[24] Varma AJ, Deshpande SV, Kennedy JF. Metal complexation by chitosan and its derivatives: A review. Carbohydrate Polymers. 2004;**55**:77-93

[25] Dutta PK, Dutta J, Chattopadhyaya MC, Tripathi VS. Chitin and chitosan: Novel biomaterials waiting for future developments. Journal of Polymer Materials. 2004;**21**:321-333

[26] Kyzas GZ, Bikiaris DN. Recent modifications of chitosan for adsorption applications: A critical and systematic review. Marine Drugs. 2015;**13**:312-337

[27] Bois L, Bonhommé A, Ribes A, Pais B, Raffin G, Tessier F. Functionalized silica for heavy metal ions adsorption. Colloids and Surfaces A: Physicochemical and Engineering Aspects. 2003;**221**:221-230

[28] Jal PK, Patel S, Mishra BK. Chemical modification of silica surface by immobilization of functional groups for extractive concentration of metal ions. Talanta. 2004;**62**:1005-1028

[29] Gandhi MR, Meenakshi S. Preparation and characterization of silica gel/chitosan composite for the removal of Cu (II) and Pb (II). International Journal of Biological Macromolecules. 2012;**50**:650-657

[30] Deng Y, Kano N, Imaizumi H. Adsorption of Cr (VI) onto hybrid membrane of Carboxymethyl chitosan and silicon dioxide. Journal of

Chemistry. 2017;**2017**:1-8. DOI: 10.1155/2017/3426923

[31] Mishima K, Du X, Sekiguchi S, Kano N. Experimental and theoretical studies on the adsorption and desorption mechanisms of chromate ions on cross-linked chitosan. Journal of Functions Biomaterials. 2017;**8**:51. DOI: 10.3390/jfb8040051

[32] Das D, Pal A. Adsolubilization phenomenon perceived in chitosan beads leading to a fast and enhanced malachite green removal. Journal of Chemical Engineering. 2016;**290**:371-380

[33] Bolto B, Gregory J. Organic polyelectrolytes in water treatment. Water Research. 2007;**41**:2301-2324. DOI: 10.1016/j.watres.2007.03.012

[34] Gregory J, Barany S. Adsorption and flocculation by polymers and polymer mixtures. Advances in Colloid and Interface Science. 2011;**169**:1-12

[35] Nasser MS, James AE. The effect of polyacrylamide charge density and molecular weight on the flocculation and sedimentation behaviour of kaolinite suspensions. Separation and Purification Technology. 2006;**52**:241-252. DOI: 10.1016/j.seppur.2006.04.005

[36] Mathieu L, Benoit B. Understanding the roles and characterizing the intrinsic properties of synthetic vs. natural polymers to improve clarification through interparticle bridging: A review. Separation and Purification Technology. 2020;**231**:115893-115917

[37] Pal P, Pal A. Surfactant-modified chitosan beads for cadmium ion adsorption. International Journal of Biological Macromolecules. 2017;**104**:1548-1555

[38] Pal P, Pal A. Treatment of real wastewater: Kinetic and thermodynamic aspects of cadmium adsorption onto surfactant-modified chitosan beads. International Journal of Biological Macromolecules. 2019;**131**:1092-1100

[39] Du X, Deng Y, Sekiguchi S, Miyamoto N, Kano N, Imaizumi H. Study on removal of Cr(VI) from aqueous solution by cross-linked chitosan. Journal of Environment Science Engineering A. 2018;**7**:1-7. DOI: 10.17265/2162-5298/2018.01.001

[40] Wu Y, Luo HJ, Wang H, Wang C, Zhang J, Zhang ZL. Adsorption of hexavalent chromium from aqueous solutions by graphene modified with Cetyltrimethylammonium bromide. Journal of Colloid and Interface Science. 2013;**394**:183-191

[41] Mahmoodi NM, Hayati B, Bahrami H, Arami M. Dye adsorption and desorption proper-ties of Mentha pulegium in single and binary systems. Journal of Applied Polymer Science. 2011;**122**:1489-1499. DOI: 10.1002/app.34235

[42] Zhao H, Liu X, Cao Z, Zhan Y, Shi X, Yang Y, et al. Adsorption behavior and mechanism of chloramphenicols, sulfonamides, and non-antibiotic pharmaceuticals on multi-walled carbon nanotubes. Journal of Hazardous Materials. 2016;**310**:235-245. DOI: 10.1016/j.jhazmat.2016.02.045

[43] Feng S, Du X, Bat-Amgalan M, Zhang H, Miyamoto N, Kano N. Adsorption of REEs from aqueous solution by EDTA-chitosan modified with zeolite imidazole framework (ZIF-8). International Journal of Molecular Sciences. 2021;**22**:3447. DOI: 10.3390/ijms22073447

[44] Montazer-Rahmati MM, Rabbani P, Abdolali A, Keshtkar AR. Kinetics and equilibrium studies on biosorption of cadmium, lead, and nickel ions from aqueous solutions by intact and chemically modified brown algae. Journal of Hazardous Materials. 2011; **185**:401-407. DOI: 10.1016/j.jhazmat.2010.09.047

[45] Dolgormaa A, Lv C-J, Li Y, Yang J, Yang J-X, Chen P, et al. Adsorption of Cu (II) and Zn(II) ions from aqueous solution by gel/PVA-modified super-paramagnetic iron oxide nanoparticles. Molecules. 2018;**23**:2982. DOI: 10.3390/molecules23112982

[46] Hubadillah SK, Othman MHD, Harun Z, Ismail AF, Rahman MA, Jaafar J. A novel green ceramic hollow fiber membrane (CHFM) derived from rice husk ash as combined adsorbent-separator for efficient heavy metals removal. Ceramics International. 2017; **43**:4716-4720. DOI: 10.1016/j.ceramint.2016.12.122

[47] Ganchimeg Y, Burmaa G, Alen S, Sukhbaatar I. Thermodynamics and kinetics of Au(III) and Ag(I) sorption on to silicon-organic polymer. Journal of Electrical Engineering. 2016;**4**:183-188. DOI: 10.17265/2328-2223/2016.04.004

[48] Lu X, Shao Y, Gao N, Ding L. Equilibrium, thermodynamic, and kinetic studies of the adsorption of 2,4-Dichlorophenoxyacetic acid from aqueous solution by MIEX resin. Journal of Chemical & Engineering Data. 2015; **60**:1259-1269. DOI: 10.1021/je500902p

[49] Du X, Kishima C, Zhang H, Miyamoto N, Kano N. Removal of chromium(VI) by chi-tosan beads modified with sodium dodecyl sulfate (SDS). Applied Sciences. 2020;**10**:4745. DOI: 10.3390/app10144745

[50] Alemu A, Lemma B, Gabbiye N, Foo KY. Adsorption of chromium (III) from aqueous solution using vesicular basalt rock. Cogent Environmental Science. 2019;**5**:1650416. DOI: 10.1080/23311843.2019.1650416

[51] Zou X, Pan J, Ou H, Wang X, Guan W, Li C, et al. Adsorptive removal of Cr(III) and Fe(III) from aqueous solution by chitosan/attapulgite composites: Equilibrium, thermodynamics and kinetics. Chemical Engineering Journal. 2011;**167**:112-121. DOI: 10.1016/j.cej.2010.12.009

[52] Debnath S, Ghosh UC. Kinetics, isotherm and thermodynamics for Cr (III) and Cr(VI) ad-sorption from aqueous solutions by crystalline hydrous titanium oxide. The Journal of Chemical Thermodynamics. 2008;**40**:67-77. DOI: 10.1016/j.jct.2007.05.014

[53] Kocaoba S. Adsorption of Cd(II), Cr (III) and Mn(II) on natural sepiolite. Desalination. 2009;**244**:24-30. DOI: 10.1016/j.desal.2008.04.033

[54] Senberber FT, Yildirim M, Mermer NK, Derun EM. Adsorption of Cr(III) from aqueous solution using borax sludge. Acta Chimica Slovenica. 2017;**64**(3):654-660. DOI: 10.17344/acsi.2017.3534

[55] Hu XJ, Wang JS, Liu YG, Li X, Zeng GM, Bao ZL, et al. Adsorption of chromium (VI) by ethylenediamine-modified cross-linked magnetic chitosan resin: Isotherms, kinetics and thermodynamics. Journal of Hazardous Materials. 2011;**185**:306-314

[56] Liu Q, Yang B, Zhang L, Huang R. Adsorptive removal of Cr (VI) from aqueous solutions by cross-linked chitosan/bentonite composite. Korean Journal of Chemical Engineering. 2015; **32**:1314-1322

[57] Zhang L, Luo H, Liu P, Fang W, Geng J. Anovelmodified graphene oxide/ chitosan composite used as an adsorbent for Cr (VI) in aqueous solutions. International Journal of Biological Macromolecules. 2016;87:586-596